Fundamentals of Speech Science

Donald J. Fucci

Ohio University

Norman J. Lass

West Virginia University

Allyn and Bacon

Boston • London • Toronto • Sydney • Tokyo • Singapore

Vice President, Education: Paul A. Smith
Executive Editor: Stephen D. Dragin
Editorial Assistant: Bridget McSweeney
Director of Education Programs: Ellen Mann Dolberg
Marketing Manager: Brad Parkins
Production Coordinator: Holly Crawford
Editorial-Production Service: Colophon
Composition Buyer: Linda Cox
Electronic Composition: Karen Mason
Cover Administrator: Jenny Hart
Cover Designer: Brian Gogolin

A Pearson Education Company
160 Gould Street
Needham Heights, MA 02494

Internet: www.abacon.com

Between the time Website information is gathered and then published, it is not unusual for some sites to have closed. Also, the transcription of URLs can result in unintended typographical errors. The publisher would appreciate notification where these occur so that they may be corrected in subsequent editions. Thank you.

Library of Congress Cataloging-in-Publication Data

Fucci, Donald J.
Fundamentals of speech science / Donald J. Fucci, Norman J. Lass.
p. cm.
Includes bibliographical references and index.
ISBN 0-13-345695-1 (alk. paper)
1. Speech. I. Lass, Norman J. II. Title.
P95.F83 1999
808.5--dc21 99-14316
CIP

Printed in the United States of America
10 9 8 7 6 5 4 3 2 1 04 03 02 01 00 99

To
Joan, Melissa, and Jessica,
and
Martha, Laura, and Jonathan

Contents

Preface • vii

Acknowledgments • ix

1 *Introduction* • 1

The Language Tool • 8
Summary • 12
Study Questions • 13
Suggested Readings • 13

2 *Anatomy and Physiology of the Speech and Hearing Mechanisms* • 15

Respiration • 15
Phonation • 24
Articulation • 37
Audition • 43
Central Nervous System • 54
Summary • 67
Study Questions • 69
Suggested Readings • 70

3 *Basic Acoustics* • 71

Sound • 71
Sinusoidal Motion • 74
Spatial Concepts • 78
Temporal Concepts • 81
Longitudinal versus Transverse Waves • 86
Sound Propagation and Interference • 88
Complex Sounds • 90
Periodicity versus Aperiodicity • 93
Sound Visualization • 100

The Decibel • 102
Summary • 106
Study Questions • 107
Suggested Readings • 108

4 *Resonance* • 109

Sympathetic Vibration • 109
Sounding-Board Effect • 111
Cavity (Acoustical) Resonance • 111
Frequency Response Curve • 121
Vocal Tract Analogy • 122
Summary • 126
Study Questions • 127
Suggested Readings • 127

5 *Acoustics of Speech Production* • 129

Acoustical Model of Speech Production • 129
Summary • 158
Study Questions • 159
Suggested Readings • 159

6 *Speech Perception* • 161

Vowel Perception • 162
Diphthong Perception • 162
Consonant Perception • 164
Suprasegmental Perception • 167
Issues in Speech Perception • 168
Other Perceptual Phenomena • 172
Theories of Speech Perception • 174
Summary • 182
Study Questions • 182
Suggested Readings • 183

Glossary • 185

Bibliography • 193

Index • 199

Preface

Fundamentals of Speech Science addresses basic concepts in speech science in an elementary, understandable manner to facilitate the learning of technical material by both undergraduate and graduate students. The book contains numerous student-friendly features, including more than 171 clear illustrations to help explain important and often very technical concepts. Other student-friendly attributes of the book include Study Questions (for review of the chapter content) and Suggested Readings (for further clarification and study), which appear at the end of each chapter. To serve as a study aid, the book also contains a glossary for important terms used throughout the book. Terms included in the glossary are in boldface type the first time they appear in the text, thereby enhancing the learning process for students. These features make *Fundamentals of Speech Science* useful not only to students in facilitating the learning process but also to instructors for explaining technical concepts; providing a source of questions and illustrations for quizzes, exams, and in-class learning exercises; and assigning additional readings on selected topics.

The book contains six comprehensive chapters. Chapter 1 (Introduction) provides an overview, including a discussion of the models used to explain how we produce and perceive the speech signal. Chapter 2 (Anatomy and Physiology of the Speech and Hearing Mechanisms) contains detailed descriptions of the anatomical systems and how they function in speech production and speech perception. Chapter 3 (Basic Acoustics) addresses basic concepts in acoustics to help the student gain an understanding of the physics of sound, which is essential to an understanding of the speech signal and the acoustics of speech production. Chapter 4 (Resonance) is concerned with the phenomenon of resonance as it relates to the articulatory aspects of speech production, another important concept for understanding the acoustics of the speech production process. Chapter 5 (Acoustics of Speech Production) contains a detailed discussion of the speech production process from an acoustical perspective to allow us to view a very complex process in a simpler and clearer manner. Chapter 6 (Speech Perception) describes the current state of knowledge of the speech perception process, also a very complex process, which is not yet fully understood. Included in this chapter are unresolved issues, perceptual phenomena, and auditory illusions that shed light on the process of speech perception, and a description of selected theories of speech perception.

Combined we have over 60 years of higher education experience (including our current teaching responsibilities) in teaching coursework in speech science and experimental phonetics at both the undergraduate and graduate levels. We have brought this experience and understanding of the learning process to our writing in this book, which is the result of a compilation and numerous revisions of material we have used in classes in packet form for a number of years. It is our intention that the contents of this volume will result in the reader's understanding of important basic concepts and current unresolved issues in speech science that will facilitate a deeper, more thorough appreciation for the complex processes of speech production and speech perception. We hope this understanding will ultimately result in increased application of concepts in speech science to clinical practice.

D.J.F.
N.J.L.

Acknowledgments

We would like to acknowledge our former mentors who provided us not only with an understanding and appreciation of the complexity of the speech production and speech perception processes but also, more importantly, with the scholarly models for us to emulate. They informed us and they inspired us by their own scholarliness and dedication. In particular, we owe a debt of gratitude to Professors Robert L. Ringel, J. Douglas Noll, Kenneth W. Burk, and Arthur S. House, who were faculty members in the Department of Audiology and Speech Sciences at Purdue University during our graduate training. In addition, Professors Ralph L. Shelton and John F. Michel, faculty members at the University of Kansas during Dr. Lass's postdoctoral studies, also served as outstanding models to emulate. The commitment of these individuals to their teaching and research has had a profound influence on our education and subsequent professional careers. We will always be indebted to them.

We would also like to acknowledge the following reviewers for their very helpful comments and suggestions: Professors Stephen Goldinger, Arizona State University; Raymond Kent, University of Wisconsin–Madison; Lauren Nelson, University of Northern Iowa; Linda Petrosino, Bowling Green State University; Jack Ryalls, University of Central Florida; and Michael K. Wynne, Indiana University School of Medicine.

And finally, to our wives, Joan and Martha, whose never-ending patience and understanding were most helpful to us throughout the duration of this writing project: our deepest gratitude and appreciation.

1

Introduction

How do we produce speech? How is speech understood by a listener? What parts of the human body are involved in speech production? What parts are involved in speech perception? These are the questions that are addressed in this book. While speech is a very complex act, to the lay observer it seems like a very simple process: You open your mouth and out come the words! Nothing could be farther from the truth, whatever that truth is. On some topics we will provide comprehensive, thorough discussions and answers to pertinent questions, while on others, the topics will simply be "addressed," because, to date, we have no definitive evidence. While we do understand some aspects of the production and perception of speech, there still remain many unanswered questions and unresolved issues.

Models of complex processes are used to better understand these processes. They provide a mechanism to observe the whole *forest* without getting caught up in the examination of individual *trees.* Communication is a complex process, and numerous models of communication can be constructed. The type of model constructed often depends on the point of view and background of those who have developed it. A psychologist may view the speech process differently than will a linguist, and a linguist may view it differently than will a speech scientist. The modeling discussed here is oriented toward, and reflects the basic terminology of, the speech and hearing sciences.

Regardless of the specific model proposed, models of speech production and speech perception must explain how a speaker can formulate a message in the brain and then convey, via the body's nervous system, appropriate impulses to the appropriate muscles of the body, which then must move the structures necessary to make the message audible, so that

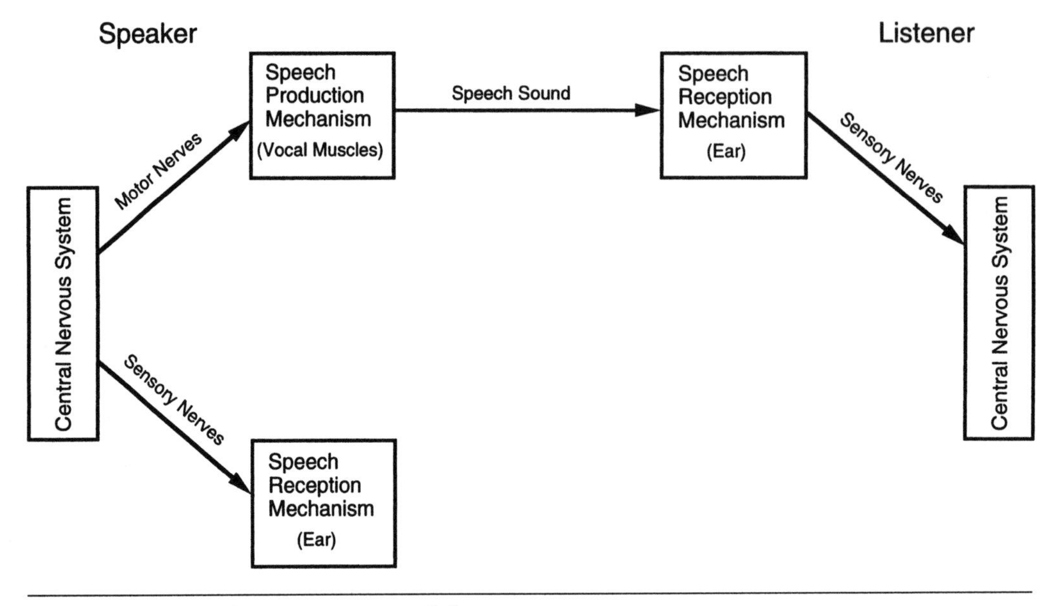

FIGURE 1-1 Speaker–Listener Model

the auditory system of the listener can interpret that message and, once it reaches the appropriate location in the nervous system, have the message interpreted and understood (Figure 1-1). Therefore, models and theories must explain the complex processes of speech production and speech perception: the acoustic signals generated by the speaker and interpreted by the listener, as well as the anatomical structures involved in speech pro-

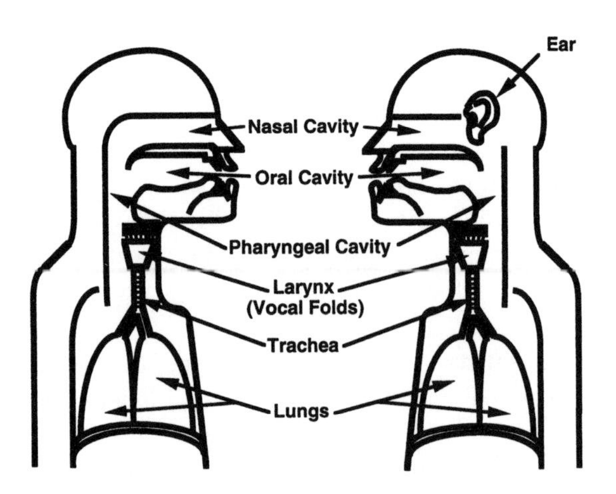

FIGURE 1-2 Speaker–Listener Process

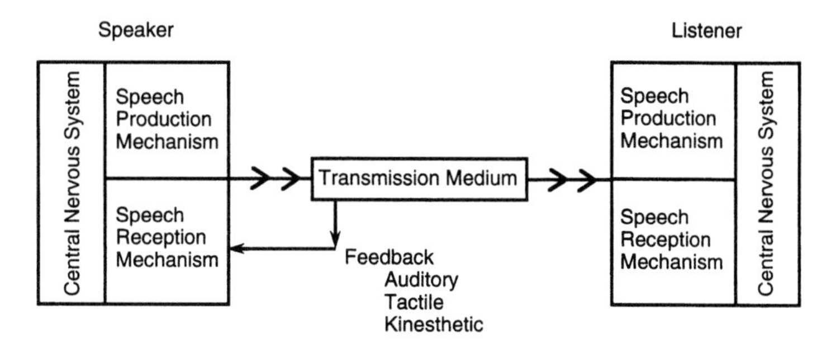

FIGURE 1-3 Speaker–Listener Model Showing Identical Speaker and Listener Components

duction (lungs, trachea, larynx, pharynx, oral cavity, and nasal cavity) and speech perception (ear, central auditory pathway, and brain) (Figure 1-2).

However, even this description is insufficient to fully explain how we produce and perceive speech. For example, how is the speaker able to produce an intelligible auditory signal when there has been damage to the brain? How is the listener able to process the incoming speech signal if there is damage to the brain or damage to the peripheral auditory system? How do speakers' regional dialects affect listeners' perceptions? How do speakers of a second language produce intelligible speech signals in that second language to listeners?

Both the speaker and listener have identical parts that are used for speech communication, so they are essentially speakers and listeners combined (Figure 1-3). The speech production mechanism has three major subdivisions: respiration, phonation, and articulation (Figure 1-4).

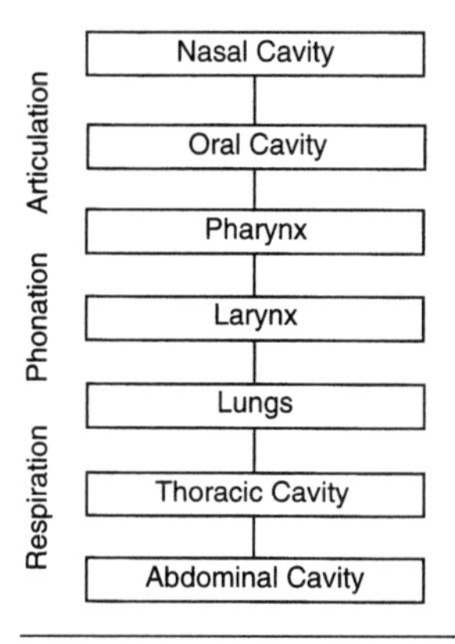

FIGURE 1-4 Anatomical Model of Speech Production

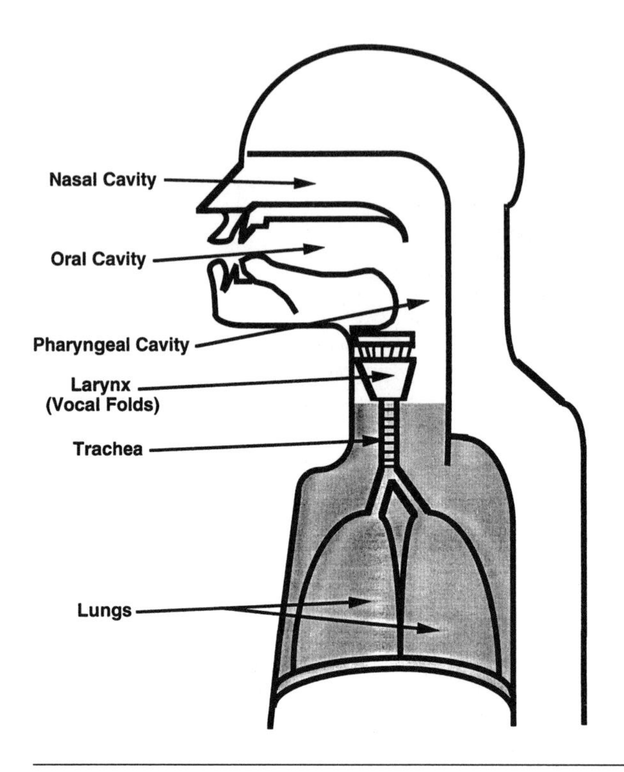

FIGURE 1-5 Speech Production: Respiration

These three components are comprised, anatomically, of the entire torso of the human body, along with some major cavities of the head.

The abdominal and thoracic regions of the body are important for the function of respiration. Respiration takes place through air exchange in the lungs (which are housed in the thorax), and this air exchange is handled smoothly by muscles of inhalation (primarily in the thoracic region) and muscles of exhalation (primarily in the abdominal region) (Figure 1-5). Speech production is accomplished on the exhalatory cycle of respiration and requires conscious control of the respiratory musculature for an extension of the exhalatory cycle over that required for normal vegetative breathing. The result needed for speech production is a steady flow of air from the lungs that can be turned into air-pulsed sound at the larynx or sound resulting from continuous or interrupted air flow at different points of constriction in the pharyngeal and oral cavities of the head.

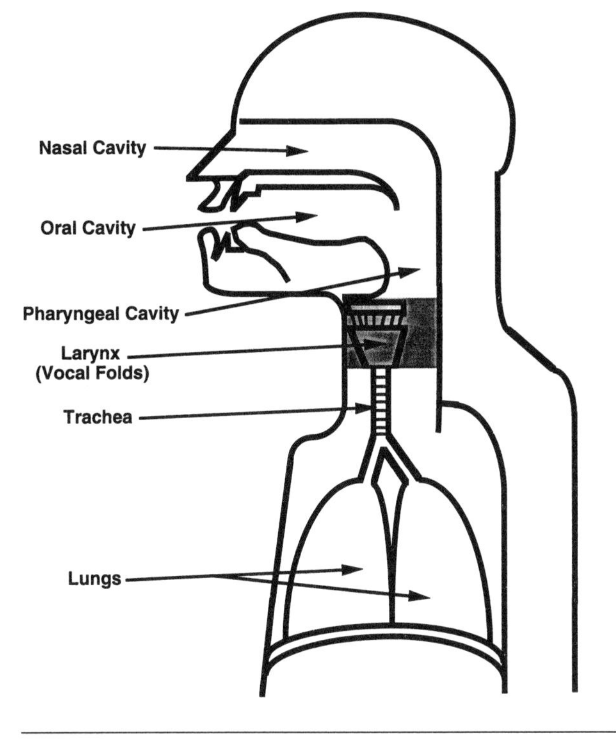

FIGURE 1-6 Speech Production: Phonation

The term *phonation* refers to the periodic impedance of the air flow coming from the lungs at the level of the larynx. The human larynx contains two muscular folds of tissue (vocal folds) that can be approximated to impede the air flowing from the lungs, causing it to be pulsed at frequencies that are high enough to be perceived as sound (Figure 1-6). The rate of vibration of the vocal folds for the average adult male is approximately 150 Hz, while the vibration rate for the average adult female is around 250 Hz.

The unique feature in the anatomical model for speech production is reflected in the human capacity for articulation. All land animals appear to have lungs for respiratory purposes, and most of them have some sort of rudimentary sound-producing mechanism, but articulation seems to be reserved for the human species. The ability to articulate allows for the formulation of speech sounds (phonemes) that are the basic building blocks of all languages. The articulators, which are comprised of structures in the pharynx, oral cavity, and nasal cavity of the

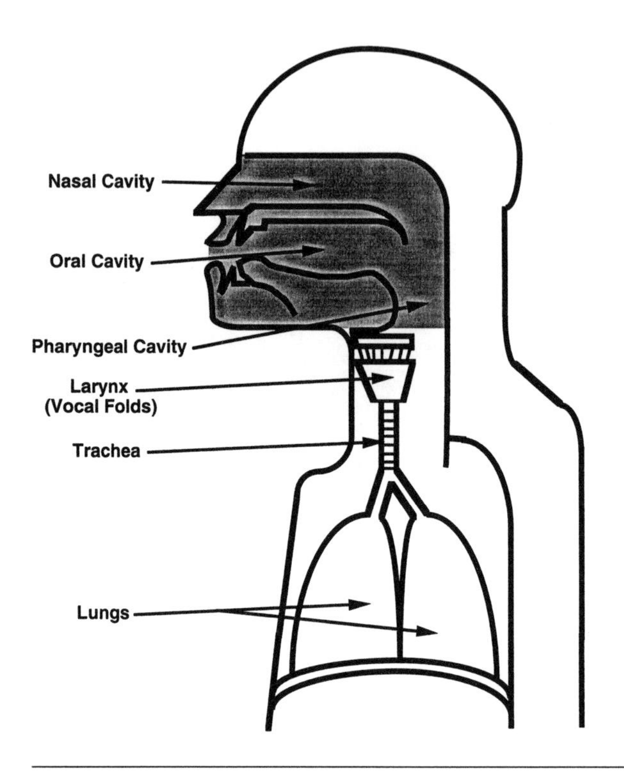

FIGURE 1-7 Speech Production: Articulation

head (Figure 1-7), are capable of "shaping" the sounds produced at the laryngeal level or at other points of constriction in the vocal tract into various speech sounds. The number of speech sounds articulated is determined by the language learned and stored in the central nervous system.

Acoustically, the speech production mechanism can be viewed as being comprised of an energy source, a voiced sound source, a voiceless sound source, and a resonance source (Figure 1-8). The energy source is the driving power needed to run both voiced and voiceless sound sources. The energy source can be viewed as steady direct-current air flow, which, when passed through the sound sources, can be converted into appropriate sound vibrations. The voiced sound source is comprised of a vibrator, which is capable of periodic vibrations needed to produce the vowels and vowel-like sounds of a language. The voiceless sound source is comprised of either narrow constrictions, which cause continuous aperiodic vibrations needed to produce the continuous consonants of a language, or complete constrictions, which can be opened quickly to cause silence, fol-

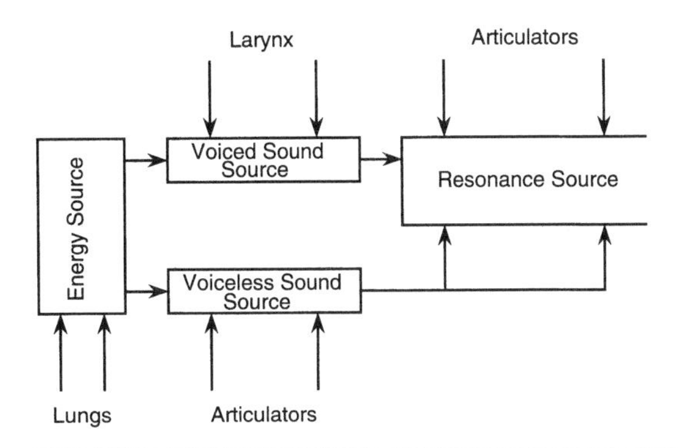

FIGURE 1-8 Acoustical Model of Speech Production

lowed by aperiodic vibrations that are characteristic of the discontinuous consonants of a language. These two sound sources also have to be able to work with each other simultaneously for the production of the voiced consonants found in languages.

The resonance source of the acoustical model of speech production is analogous to articulation. It is a simplified way of looking at the complex actions involved in articulation by reducing them to tube analogies based on dimensional and length changes of the vocal tract (pharynx and oral cavity, from the vocal folds of the larynx to the lips). If there are no constrictions in the tube and it is of a dimension and length comparable to that of the human vocal tract, voiced sound being filtered through the tube will yield a vowel close to the schwa vowel (ə) of English. However, by placing a constriction at different points along the length of the tube, different vowels and vowel-like sounds can be produced by modifying the voiced sound being filtered through it. This same kind of description, though more complex, can be used to describe consonant production as well. The number of speech sounds that the resonance source is capable of producing is determined by the language that has been learned and stored in the central nervous system.

The acoustical and anatomical models of speech production are compatible. The energy source is analogous to respiration; the voiced sound source, to phonation; and the resonance source, to articulation (Figure 1-9). However, a problem arises in regard to the voiceless sound source of the acoustical model of speech production. Anatomically, this sound-generating mechanism does not have a clear delineation in that it is actually located within the same boundaries as the structures used for

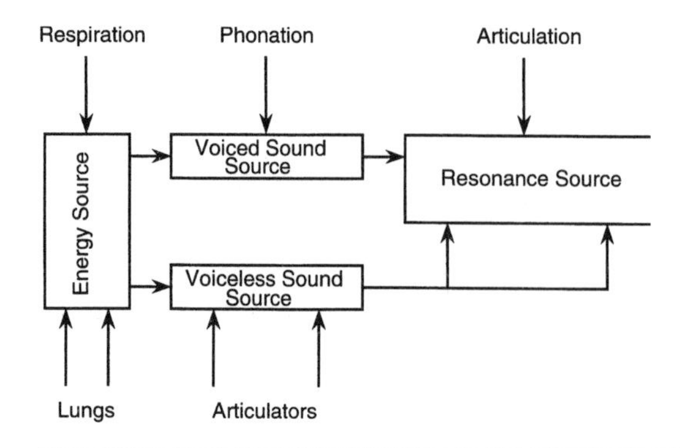

FIGURE 1-9 The Compatibility of the Acoustical and Anatomical Models of Speech Production

articulation. The assumption is that the articulatory structures can work independently and simultaneously as a voiceless sound source and as articulators. A narrow constriction between the tongue tip and alveolar ridge behind the teeth must serve both as a voiceless sound generator of continuous aperiodic vibrations and a shaping factor for the oral cavity needed to produce a particular speech sound such as the /s/ phoneme. It is easier to talk about and understand the voiceless sound source when viewing speech production from an acoustical point of view than it is to talk about it when viewing speech production from an anatomical point of view.

The Language Tool

Successful speech communication between the talker and listener is possible only if they both have the same language tool stored in their respective central nervous systems. The language tool needed for speech communication has been studied extensively but is still not completely understood. All languages have the essential building blocks of phonemes (speech sounds), lexicon (words), and syntax (rules for linking words together). These building blocks may differ in amount and kind, but they are an inherent part of language, and the human central nervous system appears capable of using them appropriately for speech communication.

There appears to be a general human propensity for language and language learning. Human children not only seem determined to achieve language skills, but they gain mastery of them as early as age 4 or 5.

Other animals are not able to master language skills in a similar fashion, even when given all possible opportunities to do so. There appears to be a "universal" factor X built into and passed on through the human genetic structure that facilitates language development.

There appears to be disagreement as to when the learning of a specific language begins. Some say that language learning begins before birth and others say that it begins at birth. In either case, it appears to be triggered by stimulation from parents or others responsible for raising a particular child. Children, through proper reinforcement, learn the language being spoken by those around them. The specific language of a child's environment is initially gained through the senses.

The senses provide the child with experiences resulting from environmental stimulation (Figure 1-10). The primary senses are vision, hearing, touch, taste, and smell, and all of these can be used by the child to some degree to gain all kinds of experiences, including those associated with language. The sense of hearing is probably the most important for initial language learning. Reading and writing are language skills that

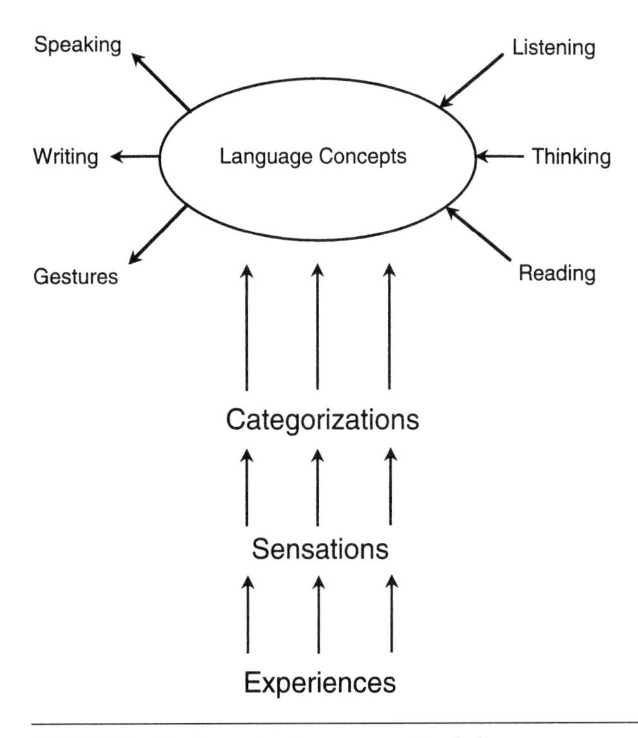

FIGURE 1-10 Speech–Language Model

are acquired after the rules of spoken language are mastered, primarily through listening to the speech of others.

The senses can become overwhelmed by too much environmental stimulation, especially if it is occurring all at once. Consequently, the child learns very early to discriminate and to categorize experiences that are being achieved through the senses. Discrimination for language learning refers to the act of building categories. Initially, the child might refer to all animals with four legs and wagging tails as dogs, especially if a dog happens to be a regular member of the child's environment. However, as the child matures, the categories are refined so that more animal categories are devised, allowing for subtle differences in animals not noticed earlier. Children with normally functioning central nervous systems appear to have little trouble categorizing things and events in their environments, and new categories are established with ease as needed.

The act of discrimination of experiences achieved through the senses leads to *language concepts* (see Figure 1-10). Language concepts are stored somewhere in the central nervous system, and there are all kinds of speculation as to how and where this storage takes place. Language concepts might be in the form of words, or they might be set up as a series of memorized pictures. They might be combinations of words and pictures, or they might take on another form not so easily defined. The actual concept of storage mode might be at the brain chemistry level or in the form of neuronal circuits. Concept storage might be in one location, such as the frontal lobes of the brain (sometimes referred to as the human lobes), or it might be spread out to all parts of the brain and controlled by some central brain mechanism.

Although there is much speculation about how the central nervous system handles language concepts, there is much less conjecture about and more understanding of how language concepts continue to be developed throughout life and how they are used in everyday situations (see Figure 1-10). The continued development of language concepts appears to occur mostly through three modes of behavior: listening, thinking, and reading. The term *listening* basically refers to hearing of speech produced by others and then internalization of language concepts conveyed through the speech of others. The traditional classroom situation of learning is based on listening as the primary way to gain new language concepts. Listening as a mode for developing language concepts is based on the notion that listeners are paying attention and that their central nervous systems are focused on what is being heard. There is also the inherent understanding that the sense of hearing rarely works alone and that information is being received through the other senses at the same time. Visual aids are also very much a part of the traditional classroom mode of learning.

Thinking as a way of developing language concepts implies that these concepts can be created internally. Creative thinking that often leads to problem solving can come from within and can lead to new language concepts that have never before been used in a routine fashion. Creative thinking is an important goal of formal education and is encouraged by teachers at all levels of the educational process.

Reading as a skill for learning and developing language concepts appears to be more difficult to master than the spoken word. Many people can neither read nor write, but they routinely use speech for communication purposes. Mastery of reading usually requires a certain amount of formal education, but once learned, reading can be a very efficient way of developing new language concepts, especially if it is something that a particular individual enjoys doing.

Language concepts can be conveyed to others through speech, writing, and gestures (see Figure 1-10). Speech is the most common tool for conveying language, mainly because it is easy to learn and efficient to use. People of all nations speak, even though the languages they use may differ. Speech is redundant, so that the message is more likely to be understood and takes little effort on the part of the speaker and the listener. An individual can talk all day long at little or no expense to the speech production mechanism. The speech reception mechanism also seems to be extremely durable and appears to be capable of processing much more than the average speaking event has to offer. Society appears to have a high regard for "good" speaking skills, and they are considered to be a critical factor for career advancement. Speech is a convenient way to influence others and to control one's own destiny. It is a universally accepted way for humans to communicate their needs and wants.

Writing is a more difficult task to master than speaking, but it is also a commonly used method for conveying language concepts. Writing skills, like reading skills, usually require some degree of formal education; writing does not seem to be the natural communication mode that speech appears to be. Speaking is preferred over writing by most individuals because speaking is easier and more efficient. Moreover, it has the added advantage of providing the listener with some idea of the emotions behind the words actually being spoken. The prosodic features of speech, such as pitch changes of the voice (intonation), inflections, stress, and juncture, are missing in writing, and they provide considerable meaning on their own. Sometimes it is not what a person says that is important, but how it is said. Writing also has certain limitations in terms of space. A person can put only so many words into a letter, whereas he can go on speaking without such restrictions.

Gestures provide a limited but useful way of communicating language concepts. One major advantage of gestures over speech is that they can be used for communication between individuals who have different native languages. It is possible to have limited success in communicating with someone who has another native language by using a variety of common gestures, but this mode of communication is not as efficient nor as fast as speech involving native speakers of the same language. Gestures seem to have a natural appeal and are often used to accompany speech output. They can be formalized into signs (like those used in sign languages) and special symbol languages used with teaching communication skills to individuals with various types of speech and language deficits. Gestures can also be expanded to encompass a holistic mode of communication referred to as "body language." The way an individual moves, sits, dresses, employs facial expressions, and uses hand signals can provide meaning as to the language concepts that person is trying to convey.

Summary

Communication is a complex process that can be viewed in a simplified manner through the use of communication models. For the communication process to be successful, it is assumed that the central nervous systems of the talker and listener are both using the same language tool for encoding and decoding messages sent between them. The language tool consists of phoneme, lexicon, and syntax structures that can vary in degree and kind for different languages. Specific languages appear to be learned through experiences derived through the senses. These experiences eventually need to be categorized into blocks that can be mastered and retained. These discriminations (categorizations) of experiences provide language concepts, which are stored somewhere in the central nervous systems of the speaker and listener. Language concepts are gained through listening, thinking, and reading, and they are conveyed through speaking, writing, and gesturing. Speech is the most common method of conveying language concepts because it is fast, efficient, and requires little effort in terms of speech production and speech reception mechanism makeup and function.

Study Questions

1. On a functional basis, the three major subdivisions of the speech production mechanism are:
   a. ______________________________
   b. ______________________________
   c. ______________________________
2. The process involving the periodic impedance of air flow from the lungs at the level of the larynx to produce voiced speech sounds is called ____________________.
3. The structures in the pharynx, oral cavity, and nasal cavity that are capable of shaping the sounds produced at the larynx (or at other points of constriction in the vocal tract) into different speech sounds are called the ____________________.
4. In the acoustical and anatomical models of speech production, the energy source is ______________________.
5. Compare and contrast the acoustical and anatomical models of speech production.
6. Language concepts are developed primarily through what three modes of behavior?
   a. ______________________________
   b. ______________________________
   c. ______________________________

Suggested Readings

Fujimura, O. (1990). Methods and goals of speech production research. *Language and Speech,* 33, 195–258.

Goldinger, S.D., Pisoni, D.B., & Luce, P.A. (1996). Speech perception and spoken word recognition: research and theory. In Lass, N.J. (Ed.), *Principles of Experimental Phonetics.* St. Louis: Mosby, pp. 277–327.

Greenberg, S. (1996). Auditory processing of speech. In Lass, N.J. (Ed.), *Principles of Experimental Phonetics.* St. Louis: Mosby, pp. 362–407.

Jusczyk, P.W. (1996). Developmental speech perception. In Lass, N.J. (Ed.), *Principles of Experimental Phonetics.* St. Louis: Mosby, pp. 328–361.

Kent, R.D., Adams, S.G., & Turner, G.S. (1996). Models of speech production. In Lass, N.J. (Ed.), *Principles of Experimental Phonetics.* St. Louis: Mosby, pp. 3–45.

Lofquist, A. (1997). Theories and models in speech production. In Hardcastle, W.J., & Lauer, J. (Eds.). *The Handbook of Phonetic Sciences.* Oxford, England: Blackwell, pp. 400–426.

2

Anatomy and Physiology of the Speech and Hearing Mechanisms

The purpose of this chapter is to provide an overview of the anatomical systems and how they function in speech production and speech perception. The speech production process can be divided into three parts: respiration, phonation, and articulation. **Respiration** provides the exhalatory air supply needed to produce speech sounds. **Phonation** is concerned with the vibratory mechanism that is needed to change the air supply into voiced speech sounds. **Articulation** is concerned with shaping sounds into specific phonemes of a language (Figure 2-1).

Respiration

For biological purposes, the respiratory process is essentially concerned with the inhalation of air for oxygen needed by the body's tissues and the exhalation of air for removal of carbon dioxide, the waste product that results from cell activity. Exhalation of air for speaking purposes is considered an overlaid function: Humans have learned to adapt a system, biologically intended to preserve the body's tissues, for purposes of communication. Respiration requires use of the bones, muscles, and tissues of the thorax and abdomen (Figure 2-2).

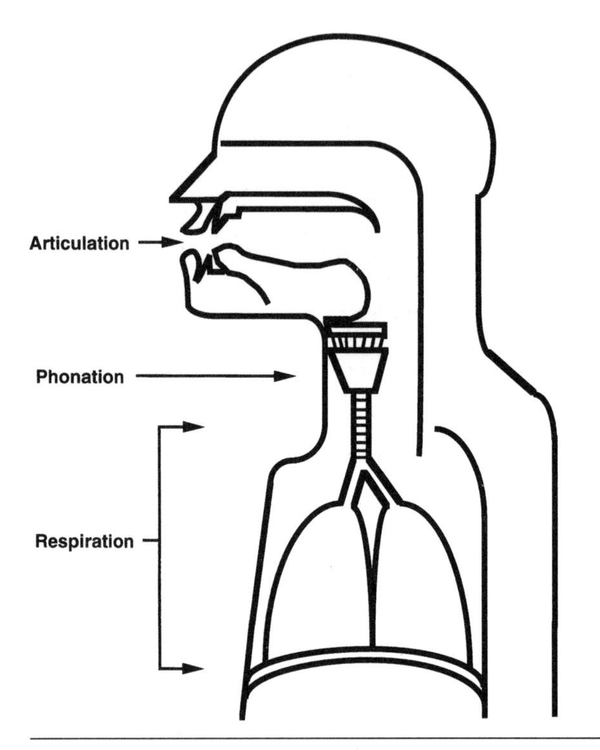

FIGURE 2-1 The Process of Speech Production

Skeletal Framework for Respiration

Vertebral Column

The **thoracic cavity** is bounded posteriorly by the **vertebral (spinal) column,** which consists of a series of small bones (*vertebrae*) that articulate with each other by means of cartilaginous discs, allowing flexibility for various human postural positions (Figure 2-3). There are 7 *cervical vertebrae* in the neck region; 12 *thoracic vertebrae* behind the heart and lungs, which are housed in the thoracic cavity; 5 *lumbar vertebrae* in the lower back region; and the *sacrum,* which consists of 3 or 4 fused *sacral vertebrae* that comprise the posterior portion of the pelvic (hip) region. There is also a *coccyx (coccygeal vertebrae)* at the lower end of the vertebral column, which comprises a rudimentary set of "tail" bones. The vertebral column serves to support the body's entire skeletal mechanism from the head above to the pelvic region below.

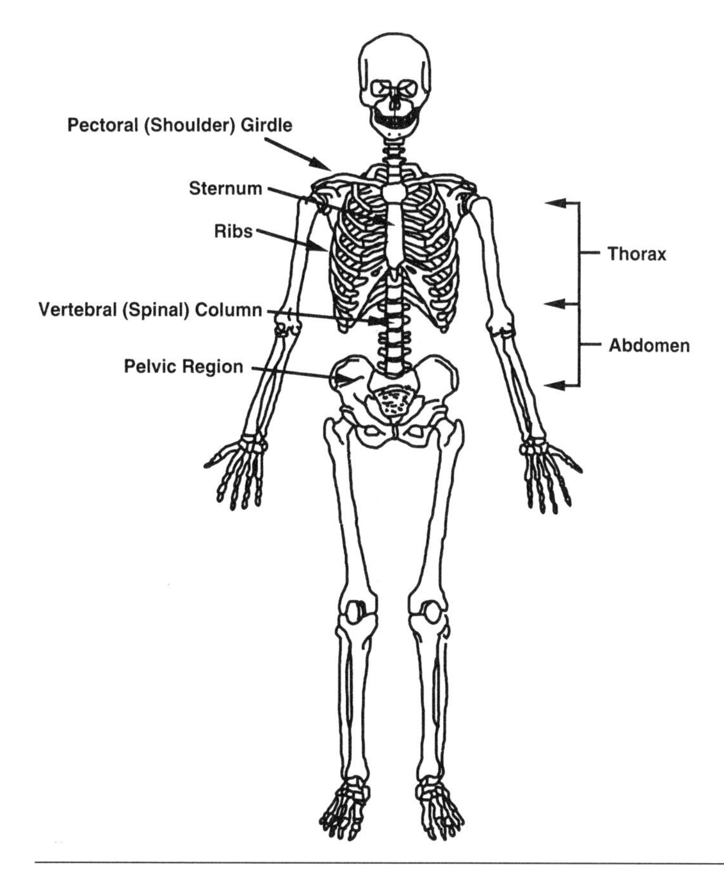

FIGURE 2-2 Anterior View of Human Skeleton

Rib Cage

The thoracic cavity is bounded laterally by the **ribs.** Humans have 12 pairs of ribs, which change in size from the first to the twelfth pair (Figure 2-4). The ribs are thin, curved shafts of bone that surround and thereby protect the lungs. Posteriorly, they attach to the thoracic vertebrae (Figure 2-5) and anteriorly, to the **sternum** (breastbone) (Figure 2-6), a fairly flat segmented bone. With the exception of the lower 2 pairs of *floating ribs,* the other 10 pairs of ribs are connected by means of their *costal cartilage* to the sternum: directly for pairs 1 through 7 (*vertebrosternal* or *true ribs*) and indirectly via a common costal cartilage for pairs 8, 9, and 10 (*vertebrochondral* or *false ribs*), thus allowing for flexibility of the

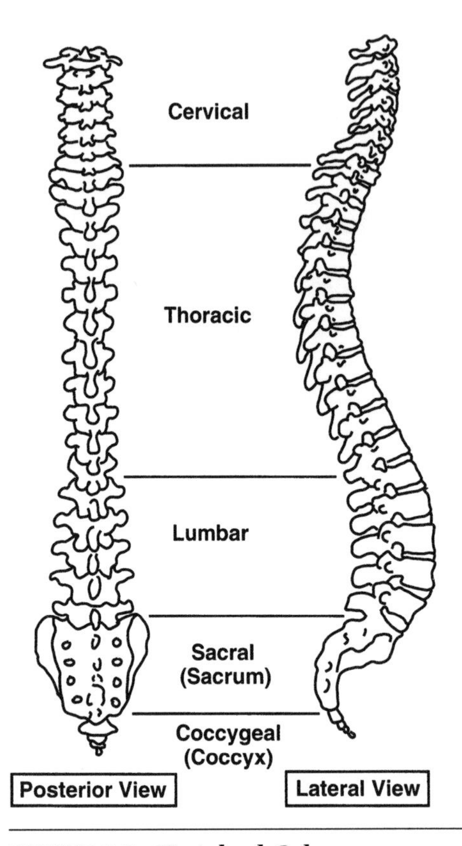

FIGURE 2-3 Vertebral Column

thoracic cage for upward and downward movement during respiration (Figures 2-4 and 2-6).

Muscles of Respiration

During the inhalation phase of respiration, muscles raise the ribs upward and outward, allowing the air sacs in the lungs to be filled with in-rushing air. During exhalation, muscles pull the ribs downward and inward, forcing air out of the lungs through the larynx, pharynx, oral cavity, and nasal cavity into the atmosphere. The thorax is separated from the abdomen by the *diaphragm* (Figure 2-7), a large dome-shaped muscle that makes up the floor of the thoracic cavity. The lateral and anterior walls of the abdomen consist of the muscles involved in exhalation.

While there are numerous muscles involved in the inhalation phase of respiration (and healthy debate as to their importance and function in the respiratory process), four appear to be of major importance in raising

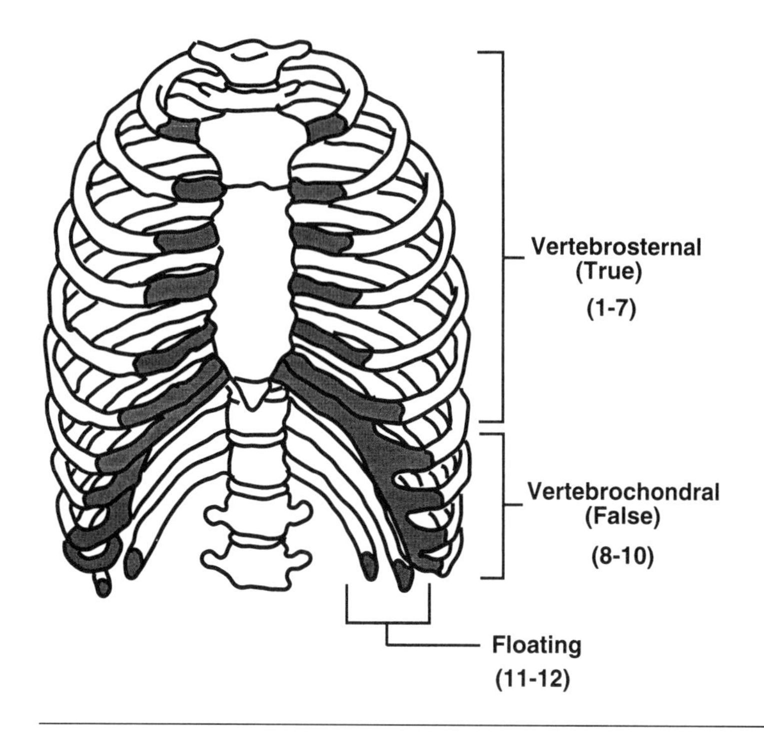

FIGURE 2-4 The Ribs

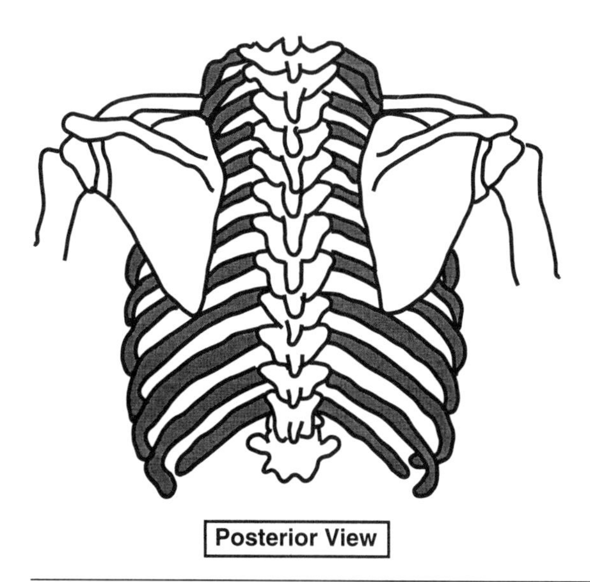

FIGURE 2-5 Ribs' Attachment to Thoracic Vertebrae

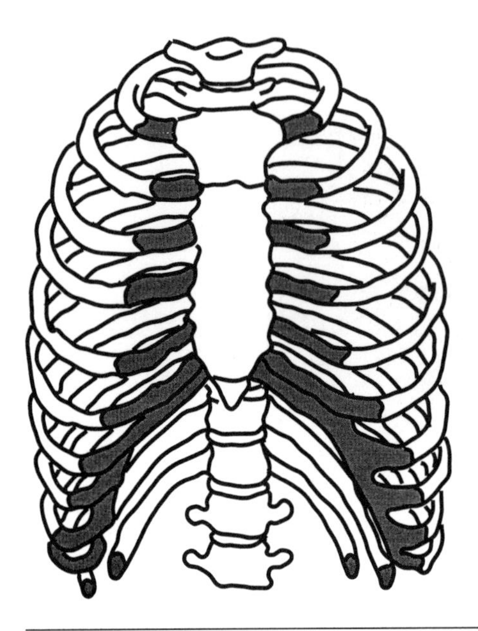

FIGURE 2-6 Ribs' Attachment to Sternum

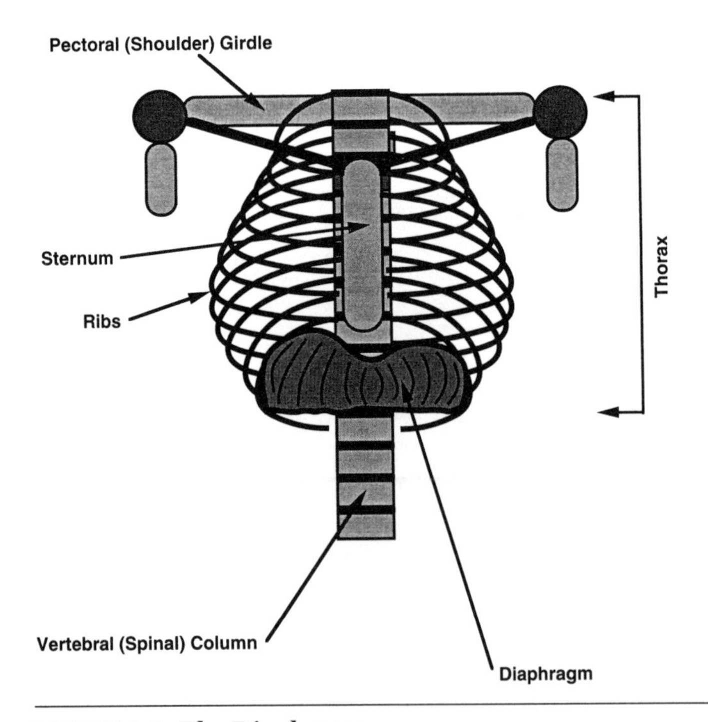

FIGURE 2-7 The Diaphragm

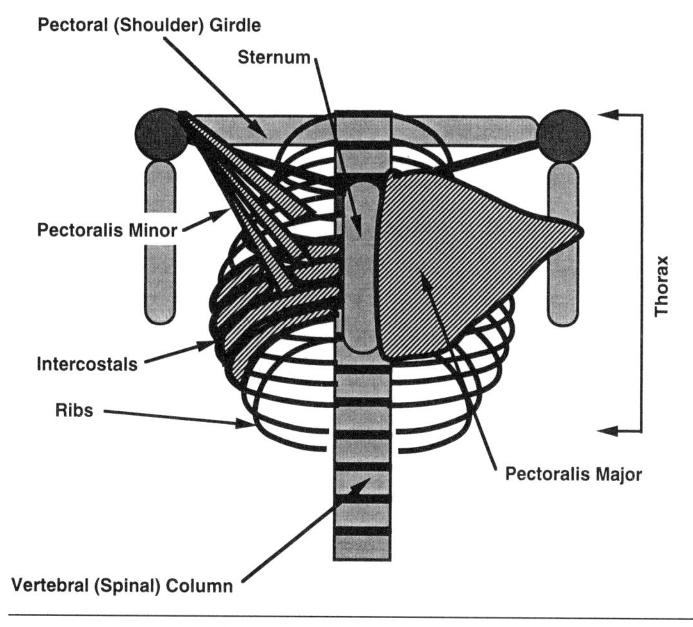

FIGURE 2-8 Muscles of Inhalation

the ribs and allowing air to rush into the lungs during this phase: the diaphragm (Figure 2-7), the pectoralis major, the pectoralis minor, and the intercostals (Figure 2-8). The *pectoralis major* is a large, powerful, fan-shaped muscle that extends from the edges of the sternum and upper ribs to the bones of the upper arm. It is located just underneath the skin and serves as a support for the breast regions of men and women. Upon contraction, it pulls the ribs upward and outward, while the arms are held fairly stationary during the contraction process.

Beneath the pectoralis major, there are three slips of muscle collectively called the *pectoralis minor*, which assists the pectoralis major because of its attachments to the middle ribs and the outer extension of the shoulder. The pectoralis major and minor work together to pull the ribs up and out during the inhalation phase of respiration.

The *intercostal muscles* fill the spaces between the ribs; when they contract, they tend to pull the ribs together, causing an overall upward movement during inhalation. These tiny muscles, which run from the bottom of one rib to the top of the next, can be divided into *external* and *internal* groups, depending on the directionality of their fibers (Figure 2-9).

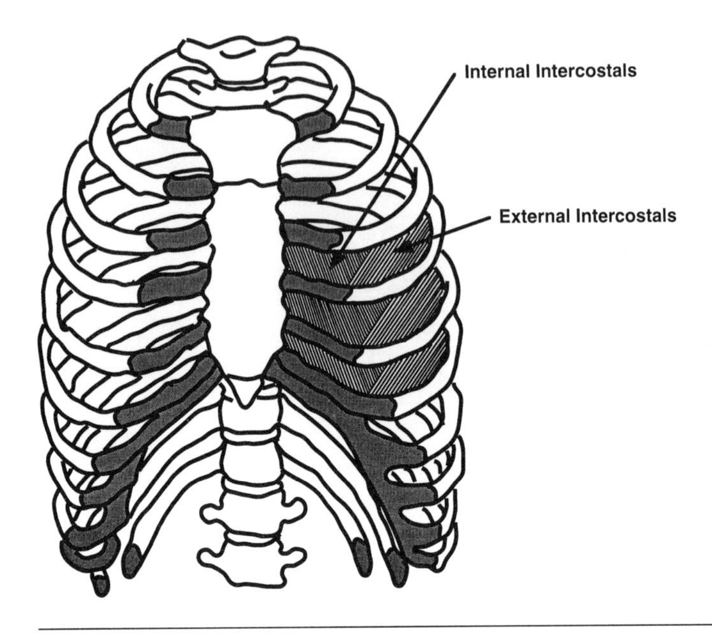

FIGURE 2-9 Internal and External Intercostal Muscles

While the pectoralis major, pectoralis minor, and intercostals raise the ribs, causing a forward and lateral extension of the volume of the thoracic cavity, the diaphragm, upon contraction, extends thoracic volume in a downward direction. When it contracts, the diaphragm flattens the floor of the thoracic cage, allowing the lungs to extend downward during inhalation.

The walls of the abdomen are comprised of four muscles involved in the exhalation phase of respiration: external oblique, internal oblique, rectus abdominis, and transversus abdominis (Figure 2-10). All of these muscles serve to pull downward on the ribs, forcing air from the lungs during the exhalation phase of respiration. The *rectus abdominis* is a strong flat muscle that runs down the front of the abdomen from the bottom of the sternum and lower ribs to the pubic region, where the paired pelvic (hip) bones meet anteriorly. The rectus abdominis is very powerful and can be used to create a strong downward force on the entire rib cage. The other muscles of exhalation are "layered" with regard to each other, forming the thickness of the side walls of the abdominal region. The *external oblique* forms the outer layer, the *internal oblique* the middle layer, and the *transversus abdominis* the inner layer of the lateral abdominal wall. The

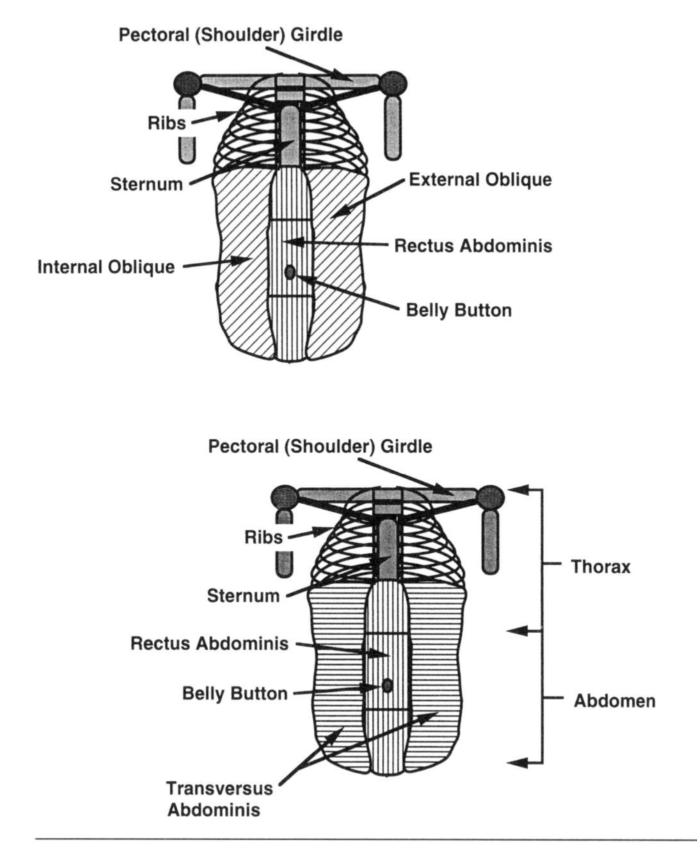

FIGURE 2-10 Muscles of Exhalation

fibers of these muscles run in different directions, adding strength to the abdominal superstructure. When they contract, they help the rectus abdominis pull the ribs in a downward direction. These muscles are critical to the exhalation phase of respiration, which is crucial for the phonatory process. They can be controlled in such a way as to provide a steady current of air to the larynx, where this current is turned into audible vibrations. Sometimes during vegetative breathing (breathing while sitting still or resting and not talking), exhalation is accomplished not by the contraction of muscles, but exclusively by the nonmuscular forces of gravity, torque (the rotational stress caused by the twisting of the ribs' costal cartilages when the ribs are raised), and elasticity of muscle and lung tissue, ribs and costal cartilages, as well as the viscera underlying the diaphragm.

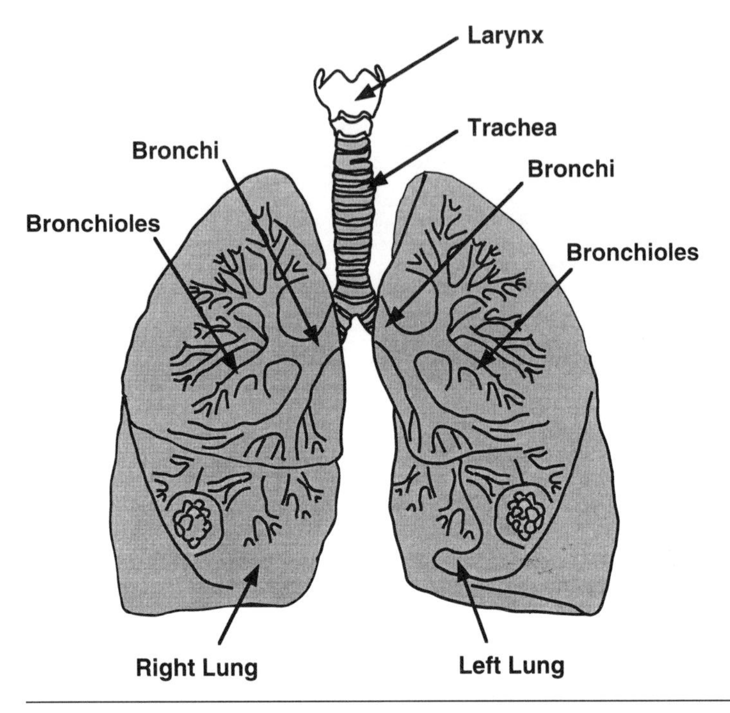

FIGURE 2-11 Trachea, Bronchi, Bronchioles, Lungs

Lungs

The **lungs** are cone-shaped and consist of millions of tiny air sacs (Figure 2-11). They adhere to the skeletal walls of the thorax and move in conjunction with them. The air sacs of the lungs feed into tubes called *bronchioles,* which in turn feed into larger tubes called *bronchi,* which finally feed into one musculo-cartilaginous tube called the **trachea.** The trachea extends upward from the lungs and lies below the larynx.

Phonation

The process of phonation occurs within the **larynx.** The larynx is a musculo-cartilaginous-membranous structure located in the anterior part of the neck. It lies below the tongue and above the lungs, to which it is connected through the trachea (Figure 2-12). The larynx is essentially a small tube that can be totally constricted at one point along its length (*vocal folds*). The primary biological function of the larynx is to protect the lungs. It has the appropriate sensory and muscular mechanisms (by

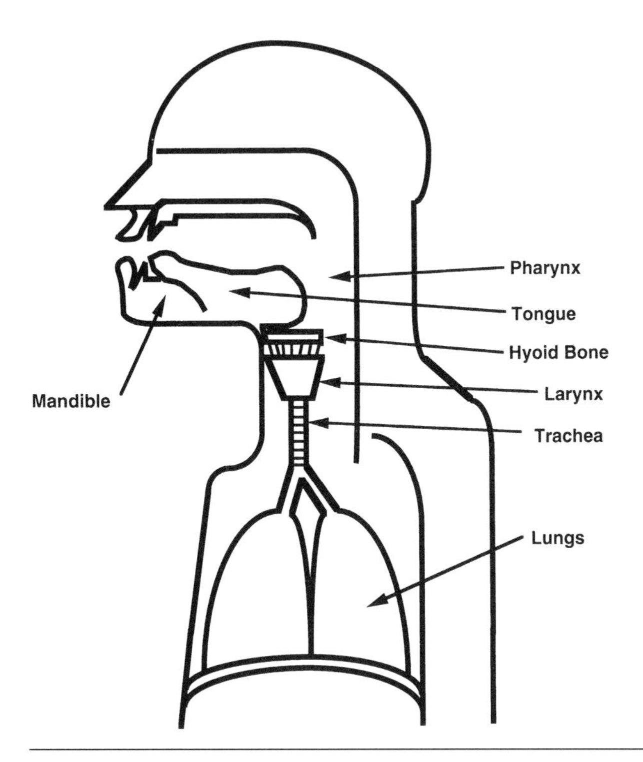

FIGURE 2-12 The Larynx

means of a rapid closing action of the vocal folds) to prevent foreign matter from getting into the lungs. The larynx is also important in the act of lifting. By trapping air in the lungs during the lifting process, it provides a solid base from which the muscles of the arms can pull. In its role in speech production, the larynx is critical to the process of phonation. It provides a smooth, rhythmic pulsation of the air emanating from the lungs, creating a periodic sound, which is identified as voicing.

The larynx is suspended from a horseshoe-shaped bone called the **hyoid bone** (Figures 2-12 and 2-13). The hyoid bone lies inside the *mandible* (lower jaw bone) and is connected to the larynx by membranous tissue, the hyothyroid membrane and lateral hyothyroid ligaments (Figure 2-14). The framework of the larynx is cartilaginous, thus providing it with the pliability needed for good voice production and overall movement during swallowing. The base of the larynx is composed of the **cricoid cartilage,** a ring-shaped structure lying directly on top of the trachea (Figure 2-15). The posterior portion of the cricoid cartilage provides

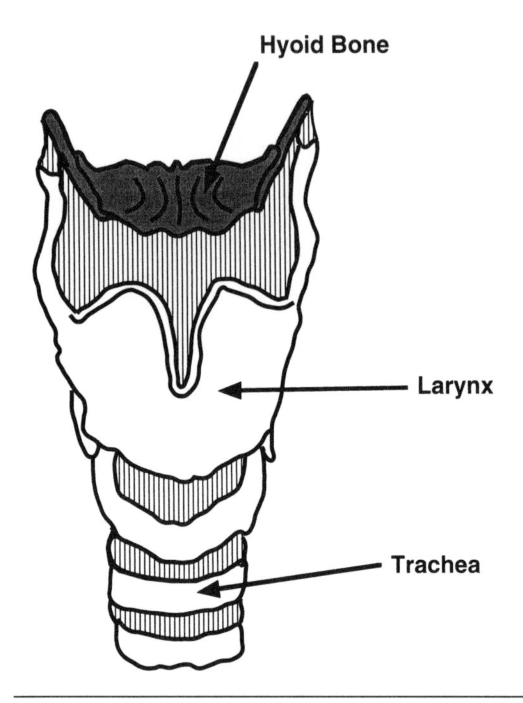

FIGURE 2-13 Anterior View of the Larynx

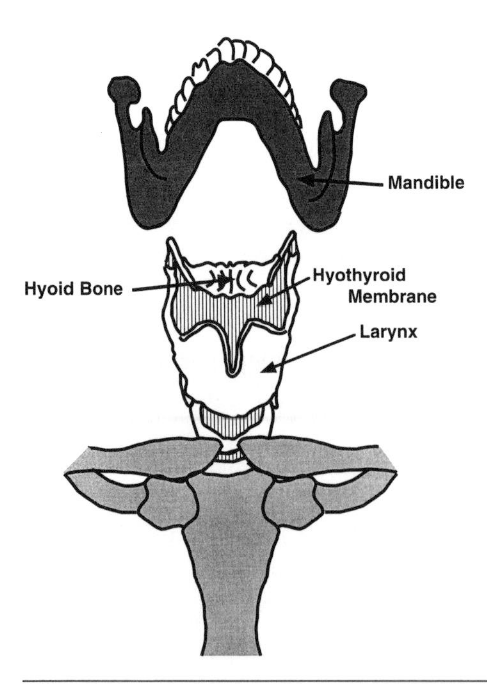

FIGURE 2-14 Membranous Connection of the Larynx to Hyoid Bone

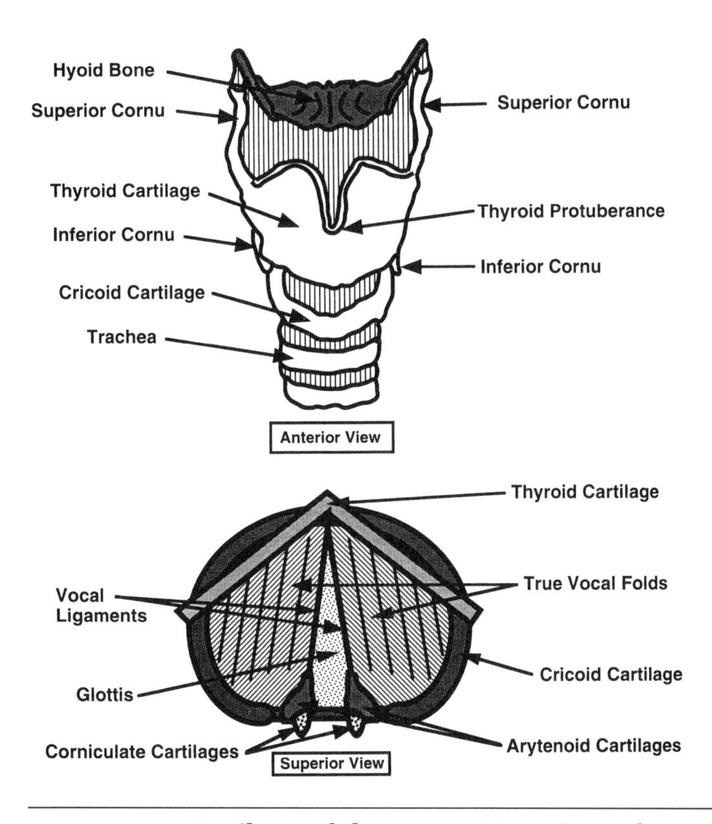

FIGURE 2-15 Cartilages of the Larynx (Anterior and Superior Views)

contact points for the two **arytenoid cartilages,** which form the anterior and posterior attachments of the vocal folds (Figure 2-16). The **thyroid cartilage** is shield-shaped and comprises the anterior aspects of the laryngeal framework (see Figures 2-15 and 2-16). This cartilage serves as a protector of the vocal folds and other interior parts of the larynx. The protuberance near the anterior–superior portion of the thyroid cartilage is called the *Adam's apple,* which appears to be more prominent in males than in females (see Figure 2-15). It is just below the thyroid protuberance that the vocal folds make their anterior attachment to the inner angle of the thyroid cartilage. The *inferior cornua* (lower horns) of the thyroid cartilage adhere to the sides of the cricoid cartilage so that these two cartilages can be rocked back and forth, causing an increase or decrease in the tension of the vocal folds (see Figure 2-15).

Posteriorly, the vocal folds are attached to the arytenoid cartilages (see Figure 2-16). These cartilages are paired and somewhat pyramidal in shape. They are located on top of the posterior portion of the cricoid car-

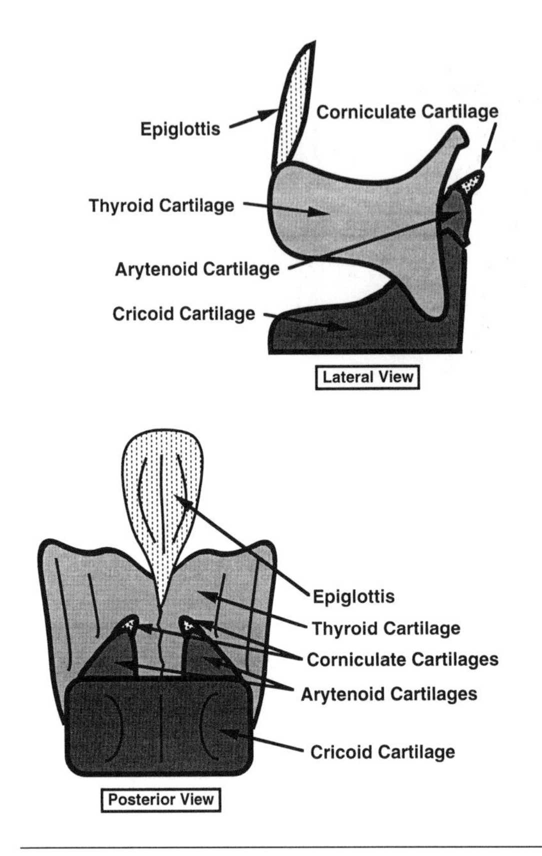

FIGURE 2-16 Cartilages of the Larynx (Posterior and Lateral Views)

tilage and can be approximated in the midline or pulled apart by appropriate intrinsic muscle activity. When approximated, the arytenoid cartilages serve to close the vocal folds (adduction) for protection of the lungs from foreign matter or for preparation of the larynx for phonatory purposes. When pulled away from each other, the arytenoid cartilages separate the vocal folds (abduction), allowing for a free respiratory passageway between the oral–nasal cavities and the lungs. The arytenoid cartilages have small upward extensions referred to as the **corniculate cartilages** (see Figures 2-15 and 2-16). The corniculate cartilages appear to aid the arytenoid cartilages in serving as an attachment for the posterior aspects of the vocal folds. The **epiglottis** is a leaf-shaped cartilage that extends upward from the thyroid cartilage to the back of the tongue (see Figure 2-16). Its function in humans is not clear. Some experts

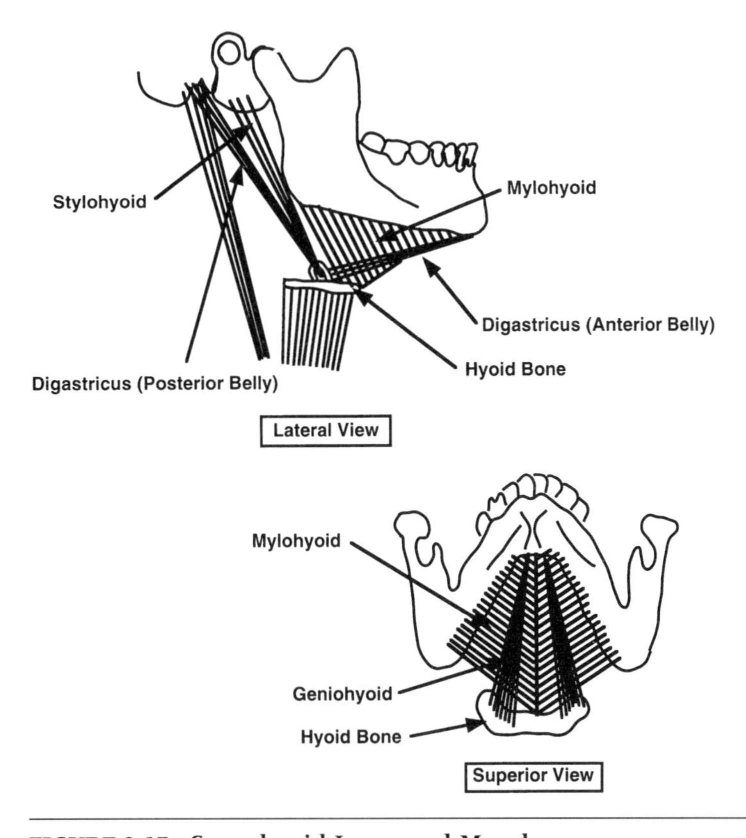

FIGURE 2-17 Suprahyoid Laryngeal Muscles

believe that the epiglottis serves to protect the laryngeal opening during swallowing by folding down over it. Others believe that it is no longer a viable part of the laryngeal framework and that at one period in evolution it was much larger, possibly extending all the way up into the nasal cavity to provide a more open passageway for breathing purposes.

The laryngeal mechanism is held in place by an extensive set of *extrinsic laryngeal muscles,* which extend from various parts of the larynx to the hyoid bone, skull, and mandible above, or to the sternum, pectoral (shoulder) girdle, and ribs below. These muscles can be classified anatomically as *supra* (above)- or *infra* (below)-*hyoid;* they can also be classified functionally as *laryngeal elevators* or *laryngeal depressors.*

Suprahyoid muscles, which serve to elevate the larynx, include the *digastricus, stylohyoid, geniohyoid,* and *mylohyoid muscles* (Figure 2-17). Of the four infrahyoid muscles, three depress the larynx: the *omohyoid, sternohyoid,*

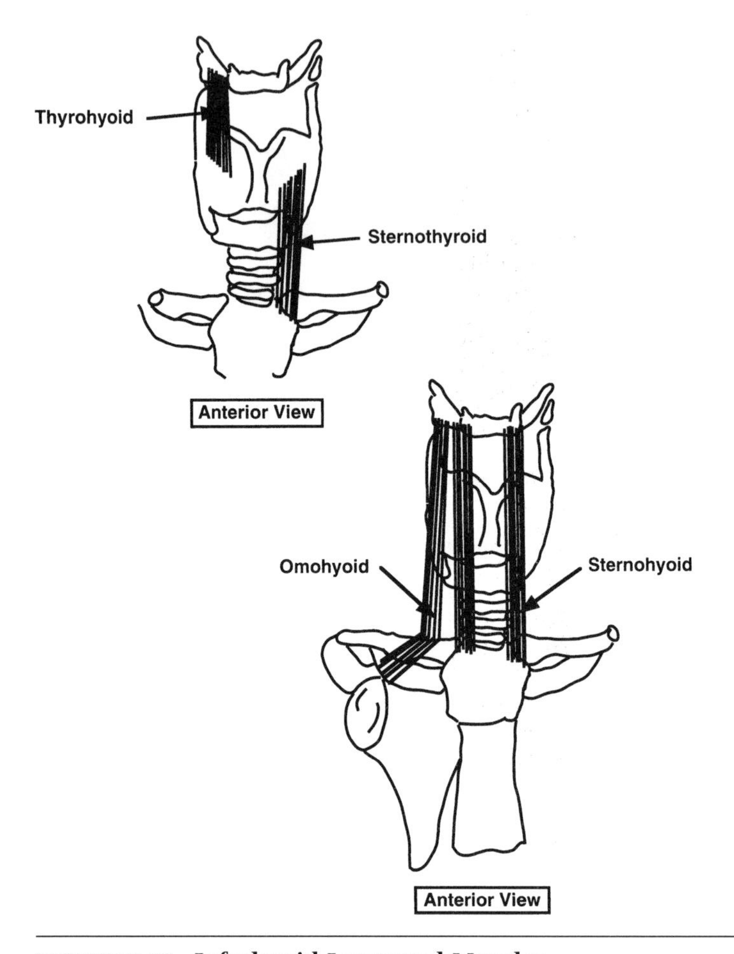

FIGURE 2-18 Infrahyoid Laryngeal Muscles

and *sternothyroid muscles* (Figure 2-18). One infrahyoid muscle, the *thyrohyoid,* is a laryngeal elevator. The extrinsic muscles are responsible for holding the larynx in the anterior midline position of the neck, while at the same time permitting it free vertical movement for phonation and swallowing purposes.

The *intrinsic laryngeal muscles* that govern vocal fold action for respiratory and phonatory purposes can be classified according to their functions: *abductor, adductor, tensor,* and *relaxer.* There is one abductor muscle, two adductor muscles, two tensors, and two relaxers. The *posterior cricoarytenoid muscle* is responsible for the opening of the vocal folds (abduction). This paired muscle runs from the back of the cricoid cartilage up to the

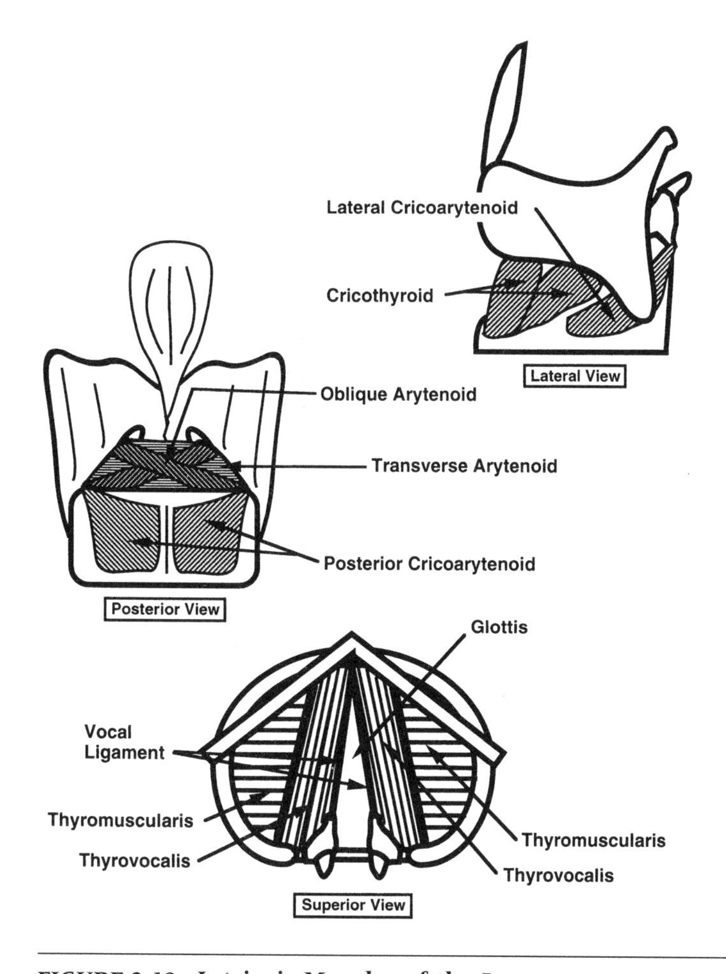

FIGURE 2-19 Intrinsic Muscles of the Larynx

side of the respective arytenoid cartilage (Figure 2-19). When contracted, the posterior cricoarytenoid muscles pull the arytenoid cartilages down and back, causing them to slide away from each other. The vocal folds are therefore spread apart, forming a triangular space between them, called the **glottis** (Figures 2-19 and 2-20).

The primary adductor muscle is the *lateral cricoarytenoid muscle.* It is also paired, and each component runs from the side of the cricoid cartilage superiorly and posteriorly to the side of a respective arytenoid cartilage (see Figure 2-19). When this muscle contracts, the arytenoid cartilages are pulled forward so that their anterior portions come together in the midline. The result is the closing of the vocal folds, which, in turn, causes the dis-

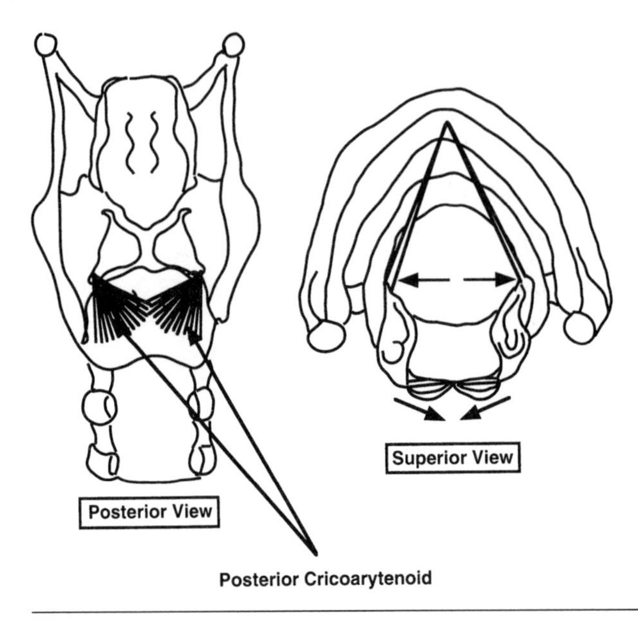

FIGURE 2-20 Posterior Cricoarytenoid Muscles and Their Actions

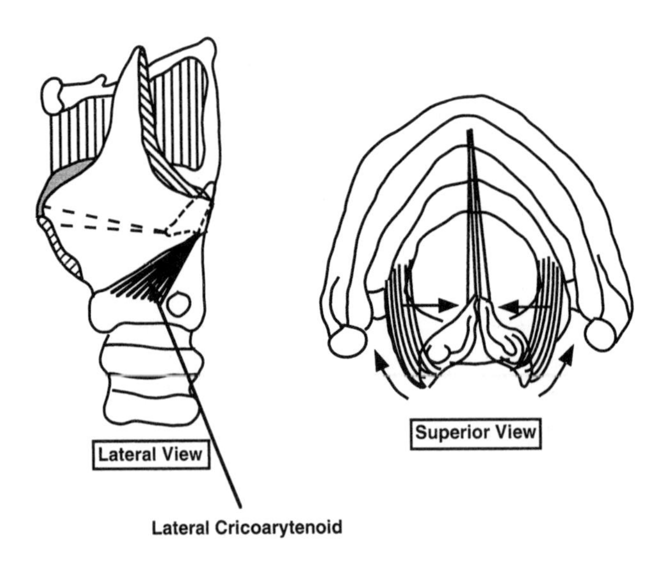

FIGURE 2-21 Lateral Cricoarytenoid Muscles and Their Actions

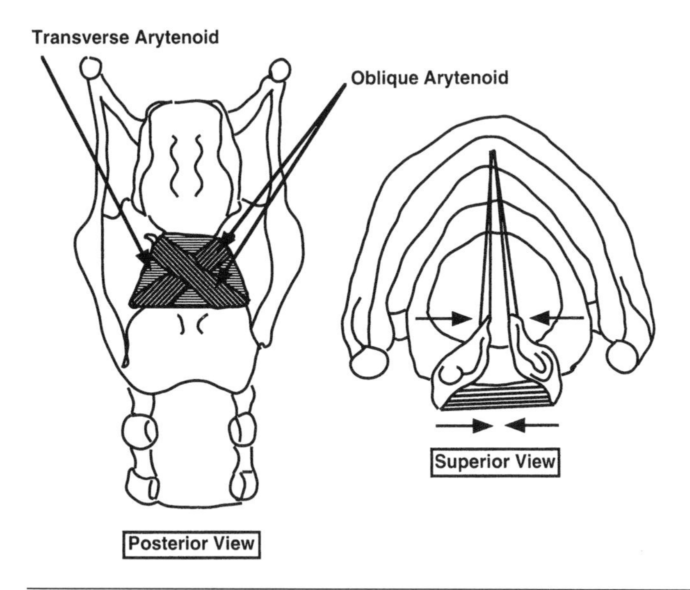

FIGURE 2-22 Transverse and Oblique Arytenoid Muscles and Their Actions

appearance of the glottis (Figure 2-21). The lateral cricoarytenoid muscle is aided in the adduction process by the *transverse arytenoid* and *oblique arytenoid muscles* (see Figure 2-19). These muscles cross between the backs of the arytenoid cartilages, and when they contract, they literally squeeze the arytenoid cartilages together in the midline position (Figure 2-22). Vocal fold tension is produced by the *cricothyroid muscle,* a paired muscle attached to the front of the cricoid cartilage that extends upward to the lower border of the thyroid cartilage (see Figure 2-19). When it contracts, the cricothyroid muscle pulls the anterior portions of the cricoid and thyroid cartilages toward each other, thereby stretching the vocal folds, which are attached anteriorly to the inner angle of the thyroid cartilage (Figure 2-23). For this action to take place, it is assumed that the arytenoid cartilages (posterior attachments for the vocal folds) are held stationary by other intrinsic muscle groups. Relaxation of the vocal folds is accomplished by some muscles that make up the folds. The vocal folds are comprised of a **vocal ligament** on their free edge, and the *thyroarytenoid muscles,* which consist of the *thyrovocalis muscle* (which runs parallel to the vocal ligament) and the *thyromuscularis muscle* (which runs parallel to the thyrovocalis muscle) (see Figure 2-19). When the thyrovocalis muscle contracts, it lengthens the vocal folds, thereby causing an increase in tension on the folds;

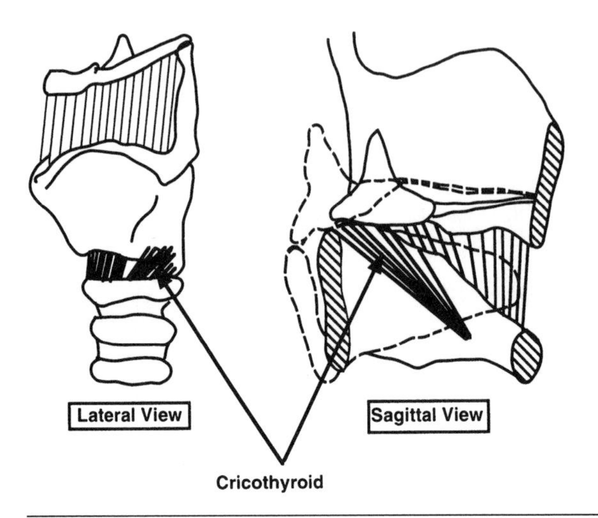

FIGURE 2-23 Cricothyroid Muscles and Their Actions

thyromuscularis muscle contraction leads to a shortening of the folds in an anterior and posterior direction, causing a lessening of the tension on the vocal folds, thus relaxing them (Figure 2-24).

In addition to cartilages and muscles, extrinsic membranes connect the larynx to outside structures (the hyoid bone above or trachea below),

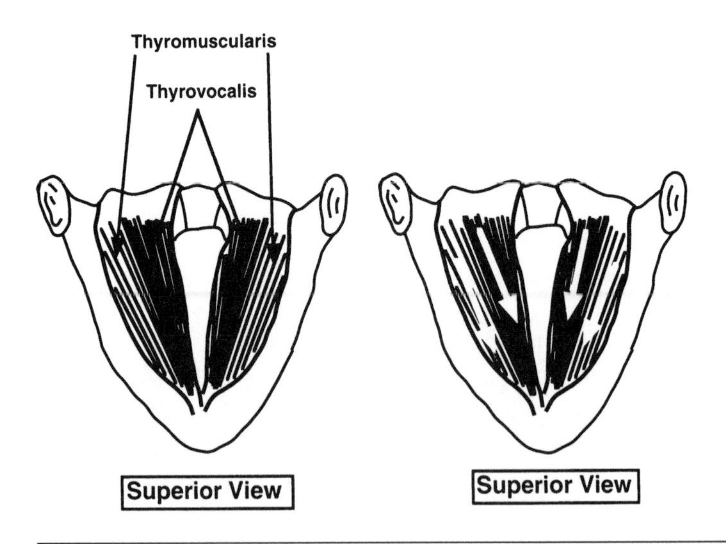

FIGURE 2-24 Thyroarytenoid Muscles and Their Actions

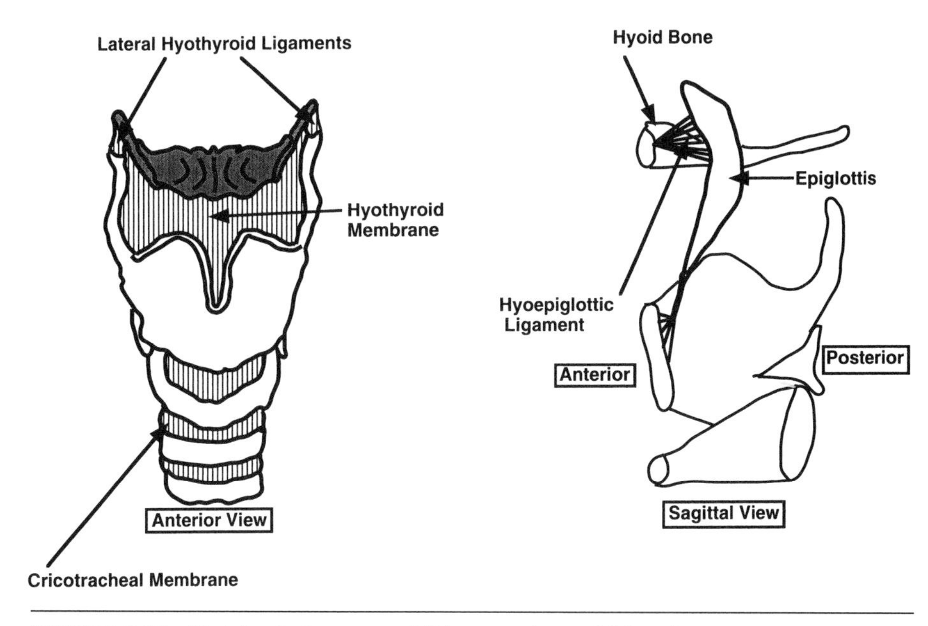

FIGURE 2-25 Extrinsic Laryngeal Ligaments and Membranes

while intrinsic membranes connect laryngeal cartilages to each other, and thereby regulate the direction as well as extent of their movements. The *extrinsic laryngeal membranes* include the unpaired *thyrohyoid* (or *hyothyroid*) *membrane,* two *lateral hyothyroid ligaments,* one *hyoepiglottic ligament,* and one *cricotracheal membrane* (Figure 2-25). The *intrinsic laryngeal membranes* are *conus elasticus* (comprised of the medial cricothyroid ligament and paired lateral cricothyroid membranes), *quadrangular membranes,* and *aryepiglottic folds* (Figure 2-26).

The larynx forms a tube that can be constricted at the level of the true vocal folds; it is an extension of the trachea below it and can be seen by means of a frontal section as the internal cavity of the larynx (Figure 2-27). The major landmarks of this laryngeal tube, in descending order, are the **vestibule** (containing the *aditus laryngis* [entranceway] and the *ventricular [false] vocal folds*), the **ventricle** (from the **false vocal folds** to the **true vocal folds,** including the **glottis,** the space between the edges of the true vocal folds), and the lower **subglottic portion** (from the *true vocal folds* to the first ring of the **trachea**).

The *ventricle* (*ventricle of Morgagni*), which separates the true from the false vocal folds, contains numerous glands that secrete mucus. The

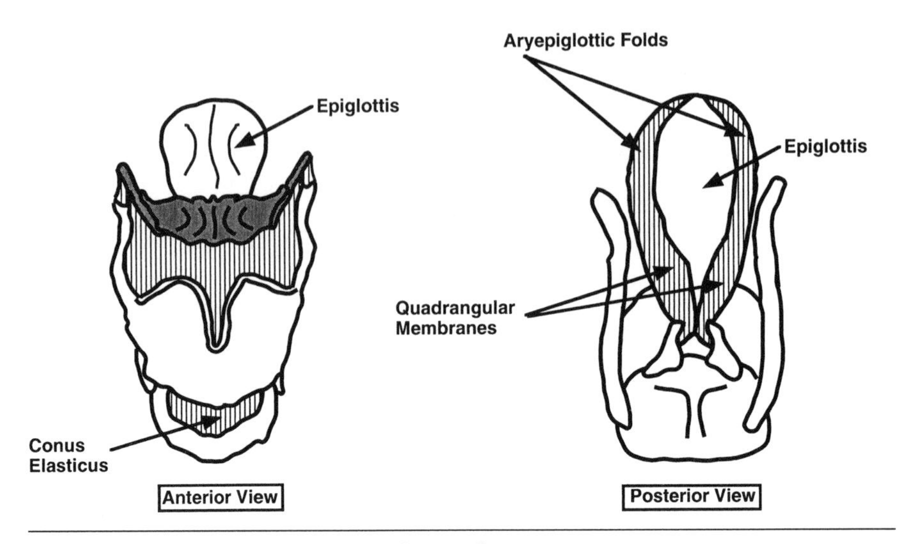

FIGURE 2-26 Intrinsic Laryngeal Membranes

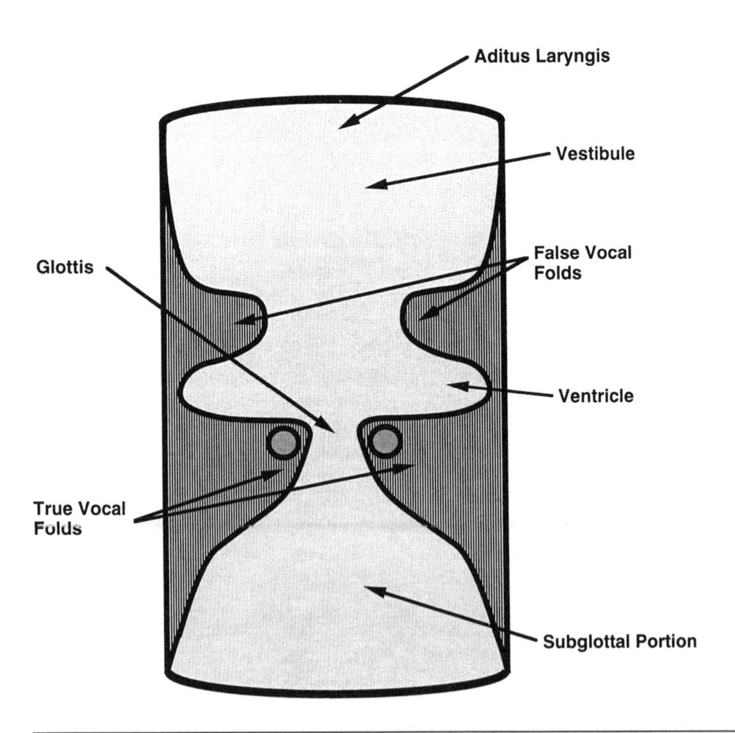

FIGURE 2-27 The Interior of the Larynx

mucus produced flows over the true vocal folds, providing them with a natural lubricant for the numerous openings and closings that they sustain during the rigorous phonatory process. The ventricular (false) vocal folds, which lie above the true vocal folds, are relatively devoid of muscle tissue and, under normal conditions, cannot be extended to a midline position for adduction purposes. The laryngeal tube is covered by a mucous membrane lining that extends from the trachea below to the tongue and lips above.

Articulation

The articulators lie above the larynx within the cavities of the pharynx (throat), oral cavity (mouth), and nasal cavity (nose) (Figure 2-28). They can change the shapes of these cavities as they approximate or come into contact with each other. The end result is the conversion of sound into speech sounds. The articulators include the lips, teeth, tongue, alveolar (gum) ridge, hard palate, soft palate (velum), and pharynx. The primary articulator is the tongue, which is critical to the production of all vowels and most consonants of English.

The **tongue** is a large muscle that makes up the floor of the oral cavity (Figures 2-28 and 2-29). The *tip* of the tongue is the thin, sharp, anterior portion, which is very important for the production of numerous sounds of English, including, for example, /t/ /d/ /n/ /s/ /z/ /l/ /θ/ and /ð/.

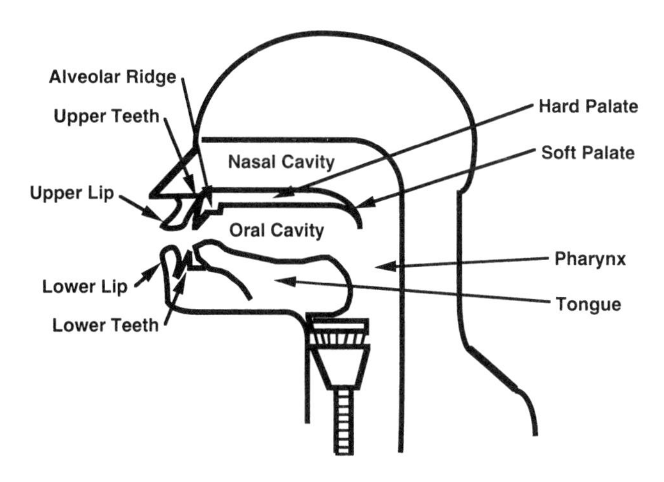

FIGURE 2-28 The Articulators

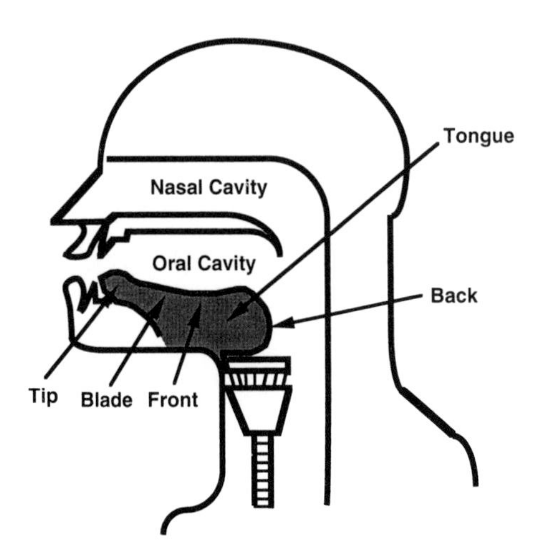

FIGURE 2-29 The Tongue

The tip appears to be able to work independently of the remainder of the tongue so that the tongue can be used to provide cavity shapes for more than one speech sound at a time. This is evident in the phenomenon of **coarticulation,** which is the simultaneous movement of two articulators. In the case of the tongue, different parts of the tongue can coarticulate to simultaneously produce two different speech sounds. The tongue can be moved forward in the oral cavity to the extent that the tip can protrude through the lips; it can also be moved backward in the oral cavity all the way to the pharyngeal wall. The tongue can be raised to touch the roof of the oral cavity, and it can be lowered to a point below the teeth. The shapes of the oral and pharyngeal cavities depend on the position of the various parts of the tongue at any given point in time.

The **lips** make up the opening that connects the oral cavity with the outside atmosphere. They represent the most anterior of the articulators (Figures 2-28 and 2-30). Human lips appear to be amenable to the articulatory process because of their pliability and mobility. The primary muscle of the lips is the unpaired *orbicularis oris muscle,* which allows the lips to provide a sphincter-like action for the production of bilabial sounds: /p/ /b/ /m/ /w/. These sounds appear to be learned early in the developmental sequence because of their visibility and the sensory readiness of the oral region shortly after birth. The lips can also be combined with the teeth for the production of the labio-dental sounds /f/ and /v/.

The **teeth** are important articulators because they serve as contact points for both the lips and the tongue (Figure 2-31). It is difficult to

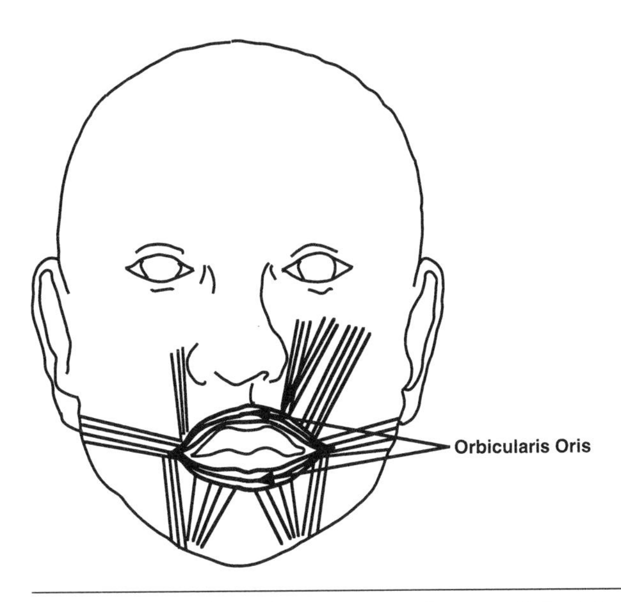

FIGURE 2-30 The Primary Musculature of the Lips

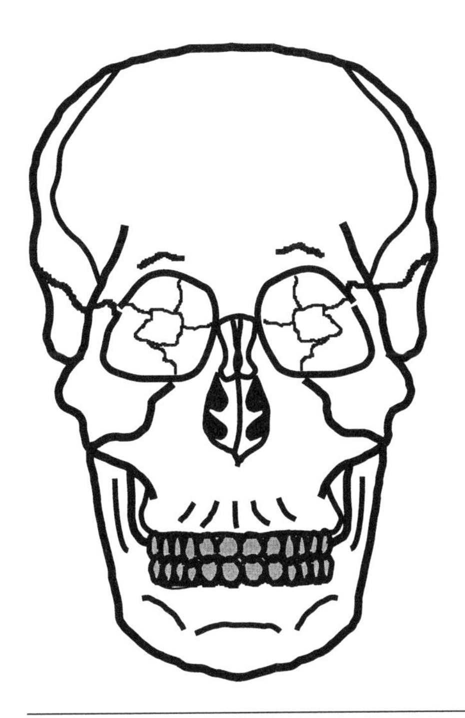

FIGURE 2-31 The Teeth

determine how important the teeth are to the articulatory process because it is known that individuals who have had their teeth removed can learn to adapt so that they can be understood by most listeners. Their /s/ sounds might not be as precise, but they are able to produce intelligible speech. Children who lose their deciduous (baby) teeth as a natural part of physical maturation sometimes exhibit articulation problems that become longstanding if they develop compensatory articulatory postures before new teeth replace the missing ones.

The **alveolar (gum) ridge** can be felt by placing the tip of the tongue just behind the upper front teeth (Figure 2-32). This ridge is a narrow bony shelf lying directly behind the upper central incisors and is an important contact point for many tongue-tip sounds (e.g., /t/ /d/ /s/ /z/ /l/ /n/). The **hard palate** extends from behind the alveolar ridge all the way back to the area of the soft palate, which is above the posterior aspects of the tongue (see Figure 2-32). The hard palate makes up the major portion of the roof of the mouth and is comprised of a mantle of bone that separates the oral cavity from the nasal cavity. It is an important contact point for lingua-palatal consonants (/ʃ/ /ʒ/). In the case of vowels, the tongue does not come into actual contact with the alveolar or palatal regions, but instead forms a narrow opening to provide the appropriate cavity shape for the particular vowel being produced.

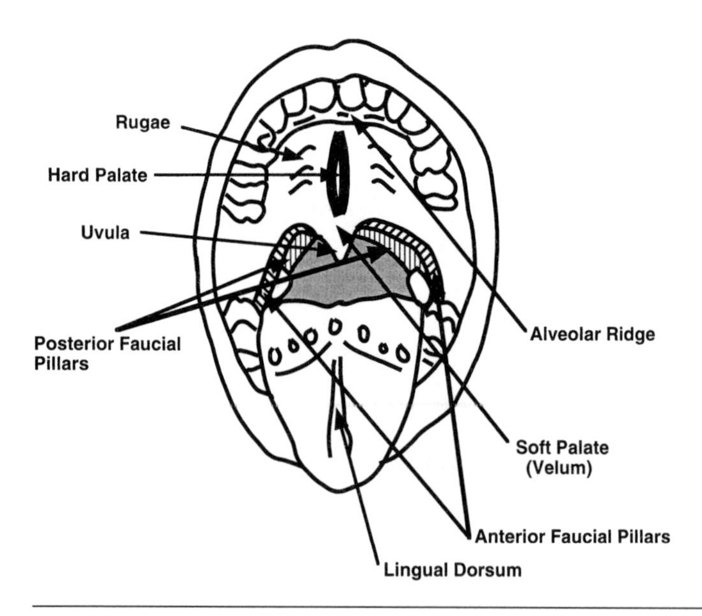

FIGURE 2-32 Oral Cavity and Adjacent Structures

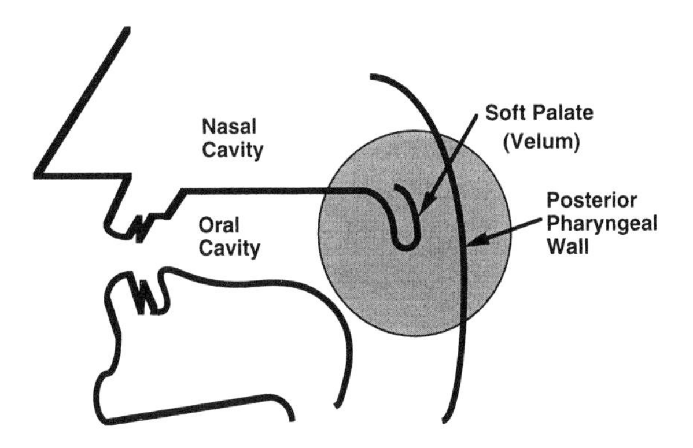

FIGURE 2-33 Velopharyngeal Port Mechanism (Lowered Velum)

The **velum (soft palate)** is a muscular protrusion that extends backward from the hard palate (see Figure 2-32). During respiration, its free end (*uvula*) hangs down in the back of the oral cavity so that the nasal cavity has unobstructed access to the lungs (Figure 2-33). For the production of all but the three nasal speech sounds (/m/ /n/ /ŋ/), the soft palate is raised upward to make contact with the posterior pharyngeal wall so that the nasal cavity is separated from the oral and pharyngeal cavities (Figure 2-34). This contact point is referred to as the **velopharyngeal port mechanism,** and the result is called **velopharyngeal closure** (velo = soft palate; pharyngeal = pharynx). During the production of nasal speech sounds, the velum is lowered, providing a nasal coupling with the oral and pharyngeal cavities (see Figures 2-28 and 2-33). The velopharyngeal port mechanism is frequently a problem area for individuals with a cleft of the palate. Because of the misalignment of the roof of the oral cavity, the velum will not function properly in conjunction with the pharynx in order to close the nasal cavity completely from the oral and pharyngeal cavities. The result is excessive nasality (hypernasality) and nasal emission of air for nonnasal speech sounds (causing a distortion of these sounds).

The **pharynx** (throat) is a cavity lying behind the nasal and oral cavities and upper larynx. It serves as an important link between numerous other cavities: as a connector between the oral cavity, nasal cavity, and larynx for speech production purposes (Figure 2-35). The pharynx also connects the oral cavity with the esophagus, the tube located behind the trachea through which food passes before entering the stomach. The pharynx connects the nasal and oral cavities to the larynx for respiratory

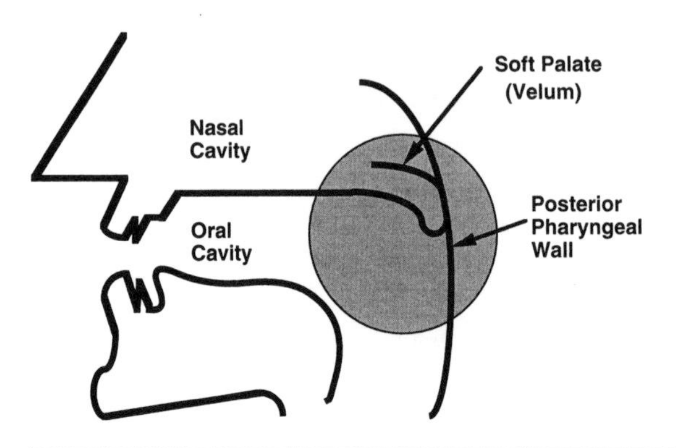

FIGURE 2-34 **Velopharyngeal Port Mechanism (Elevated Velum)**

purposes. In addition, it connects the nasal cavity to the ear (Figure 2-36), so that atmospheric air pressure can be equalized with pressure in the middle ear cavity. When discussing speech articulation from an acoustical point of view, the pharynx is considered a part of the vocal

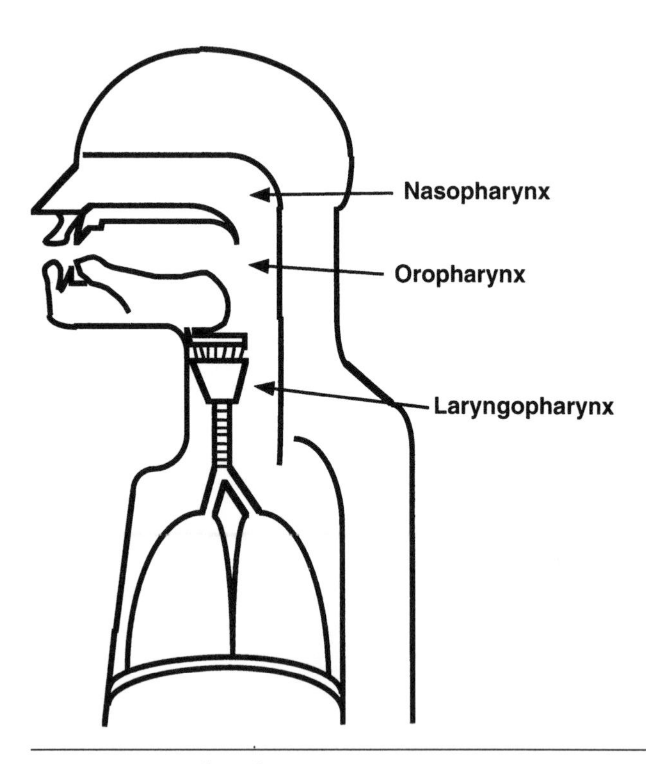

FIGURE 2-35 **The Pharynx**

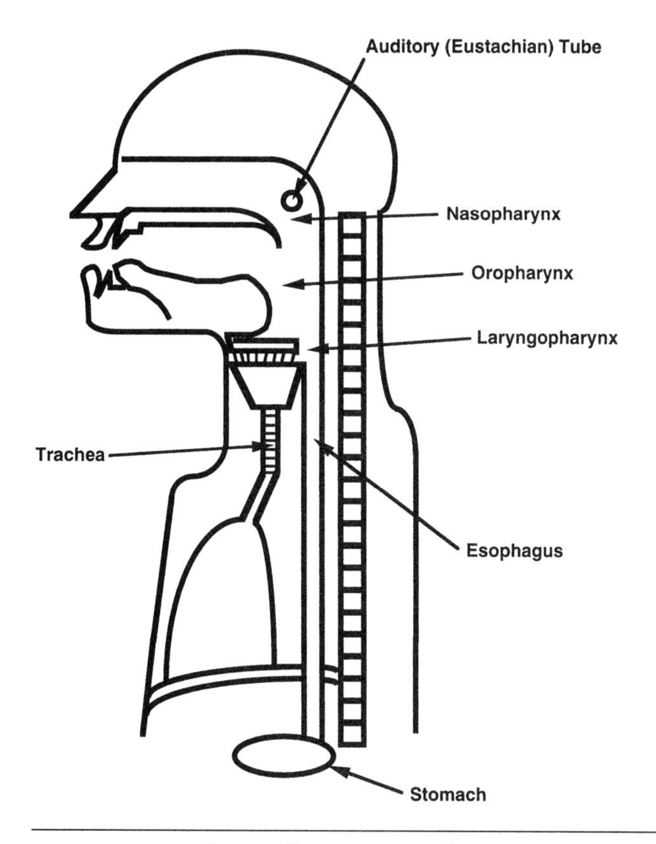

FIGURE 2-36 **The Auditory (Eustachian) Tube**

tract that extends from the glottis of the larynx to the lips. The pharynx serves not only as a contact point for the posterior aspects of the tongue during articulation, but also as a critical part of the velopharyngeal port mechanism, which is essential for the regulation of nasal cavity participation during the production of various speech sounds.

Audition

The hearing mechanism is critical to the processes of speech production and speech perception. It is responsible for the initial learning of speech from others, the monitoring of a speaker's own speech (feedback), and the reception of the acoustic speech signal for continued communication purposes throughout life. The anatomical structure of the hearing mech-

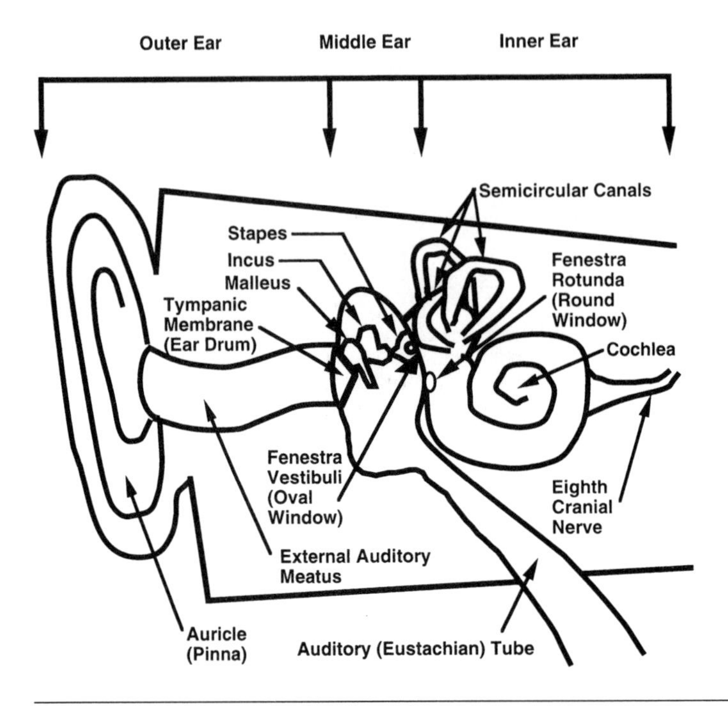

FIGURE 2-37 The Hearing Mechanism

anism can be divided into three basic parts: the outer ear, the middle ear, and the inner ear (Figure 2-37).

Outer Ear

The **outer ear** consists of the *auricle* (or *pinna*) and the *external auditory meatus* (see Figure 2-37). The auricle (pinna) is that part of the ear that can be seen protruding from the side of the head (Figure 2-38). In humans it has an irregular shape and is comprised of cartilage and soft tissue that forms a "cup" around the entrance to the external auditory meatus. Although the auricle may serve to help trap and direct sound into the external auditory meatus in numerous species, its function is somewhat limited in that capacity in humans. Unlike the auricles of other animals, such as horses and hippopotamuses, as well as domesticated dogs and cats, usually the human auricle cannot be directed toward a sound source independently of the head.

The external auditory meatus (outer ear canal) is very narrow and serves as a channel leading to the tympanic membrane (eardrum), which is located at its medial (inner) end (Figure 2-39). The external auditory

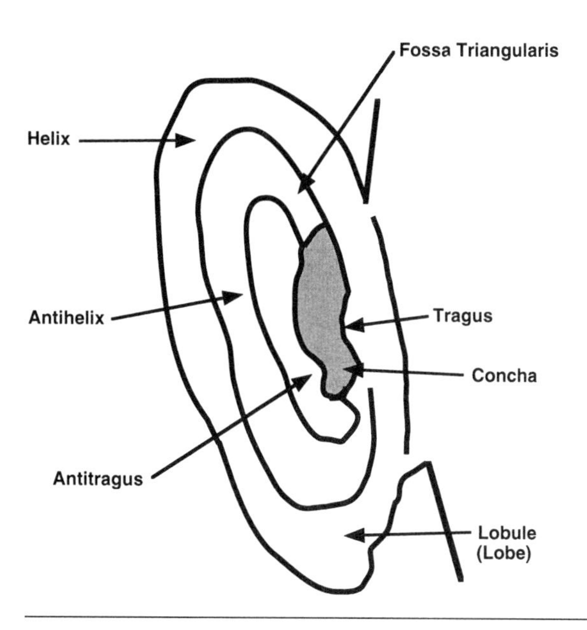

FIGURE 2-38 The Auricle (Pinna)

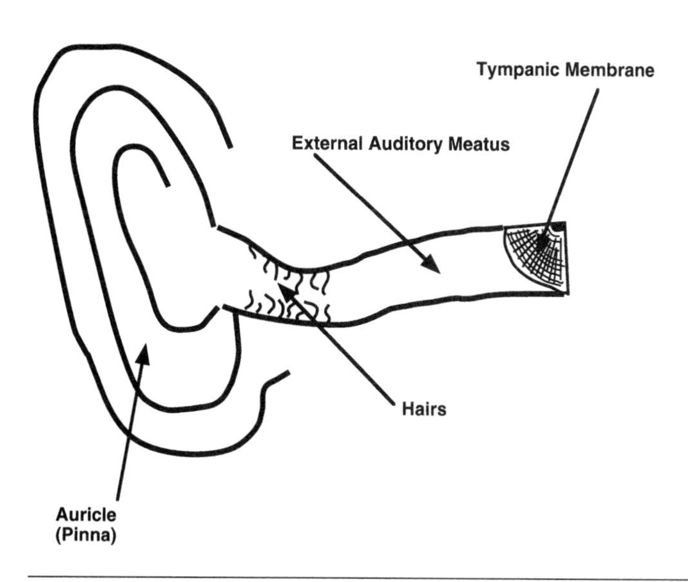

FIGURE 2-39 The External Auditory Meatus

meatus serves to protect the tympanic membrane from the outside atmosphere's foreign elements. Usually, the canal is somewhat S-shaped; consequently, the tympanic membrane cannot be seen by looking into the canal unless the auricle is first pulled up and back so that the canal is straightened. Glandular tissue located within the walls of the external auditory meatus produces a wax (lubricant) called **cerumen,** which is designed to keep the canal supple and clean. The portion of the canal nearest the auricle is also lined with hairs, which provide more protection for the tympanic membrane.

The outer ear mechanism (auricle and external auditory meatus) is designed to receive airborne sound and to serve as a channel for that sound to reach the tympanic membrane (see Figure 2-37). Once the acoustic signal reaches this membrane, the sound is converted into mechanical vibrations of the membrane itself. The tympanic membrane is a delicate structure that seals off the medial (inner) end of the external auditory meatus. It is cone-shaped and well suited to trapping the sound disturbances reaching it. The membrane serves as the partition between the outer and middle ears (see Figure 2-37). Its major parts are shown in Figure 2-40. The *pars tensa* comprises the largest portion of the tympanic membrane, containing numerous fibers that contribute to the taut nature of this portion of the membrane. In contrast, the *pars flaccida* is a small triangular area on the superior portion of the membrane with very few

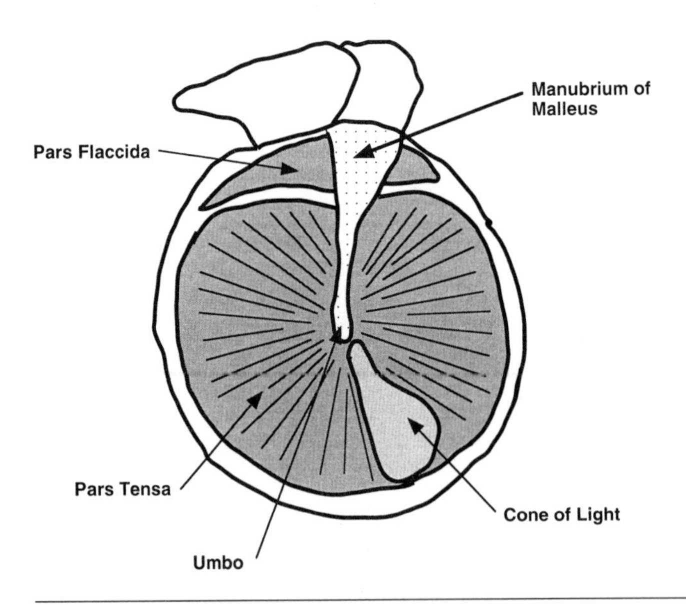

FIGURE 2-40 The Tympanic Membrane

fibers and is therefore flaccid in nature. The *umbo* is the center point of the tympanic membrane (representing the projection from the *manubrium* of the *malleus*) from which the *cone of light* radiates when the membrane is viewed via an *otoscope*, a device used to examine the outer ear.

Middle Ear

Once the airborne sound reaches the tympanic membrane and is converted to mechanical energy by setting the membrane into vibration, it remains in a mechanical mode as it is transmitted through the **middle ear** mechanism. The middle ear is a hollow cavity located behind the tympanic membrane and divided into the *tympanic cavity proper* (or *tympanum*) and the *epitympanic recess* (or *attic*) (Figure 2-41). It is filled with air supplied by the *auditory* (*Eustachian*) *tube* (Figures 2-37 and 2-41), which runs from the tympanic cavity to the back of the pharynx at a location just above the velum.

The middle ear serves as a means for equalizing the air pressure on the inside of the tympanic membrane with the air pressure within the external auditory meatus so that the membrane does not burst inward.

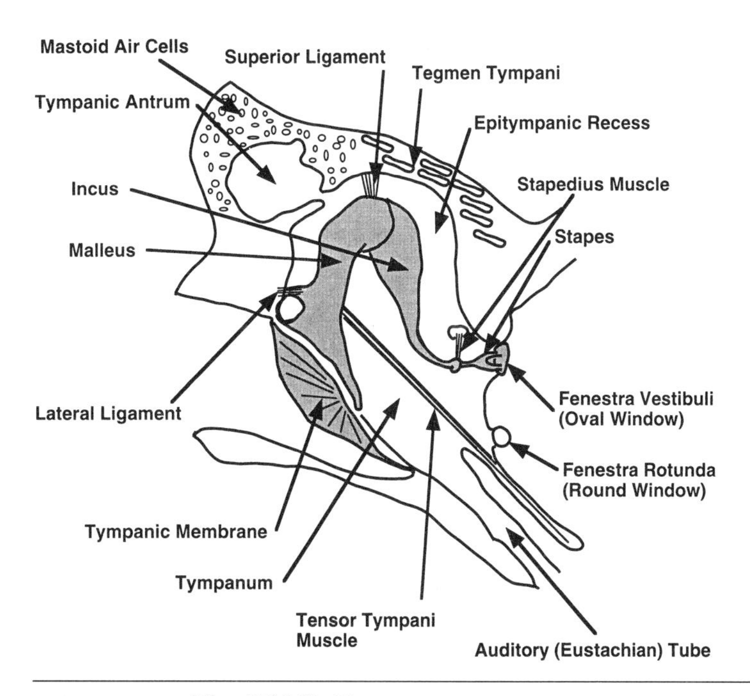

FIGURE 2-41 The Middle Ear

A bridge of three bones called the *ossicles* (or *ossicular chain*) is formed across the middle ear cavity (see Figure 2-41) for the purpose of transmitting the vibrations of the membrane to the inner ear mechanism where the sensory end organ of hearing is located. The first ossicle is called the *malleus* and is directly attached to the tympanic membrane. The second ossicle is the *incus,* which is attached to the malleus. The *stapes,* the third bone, is attached to the incus. Its footplate is inserted into the *fenestra vestibuli* (*oval window*), which connects the ossicular chain to the inner ear mechanism. The ossicles' Latin names reflect the structures that they resemble: *malleus,* hammer; *incus,* anvil (as used by blacksmiths); *stapes,* stirrup (like the stirrup on a saddle).

The ossicles are held in place by ligaments and two tiny muscles (*tensor tympani* and *stapedius*) (see Figure 2-41). The two muscles can tighten the ossicular chain in the event of loud noise exposure and therefore provide some protection to the inner ear mechanism against very intense sounds. Normally, the ossicles vibrate in tune with the tympanic membrane and deliver the vibrations of the membrane faithfully into the inner ear mechanism.

Inner Ear

The **inner ear** mechanism consists of two major parts: vestibular and cochlear (Figures 2-37 and 2-42). The *vestibular apparatus* is concerned with the sense of balance and spatial orientation. It consists of three *semi-*

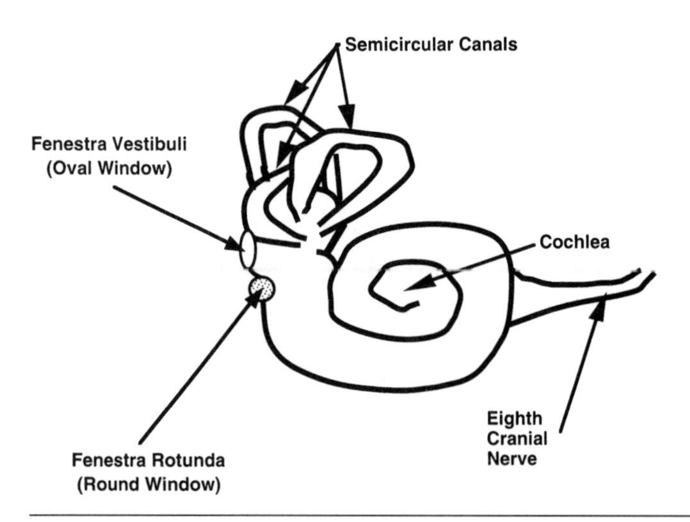

FIGURE 2-42 The Inner Ear

circular canals (*superior, lateral,* and *posterior*), each of which represents a body plane in space. The canals are filled with a fluid that moves in conjunction with head and body activity. The fluid movement within the semicircular canals is registered in various parts of the brain, which can then cause appropriate muscle–motor signals to occur in order to maintain proper body posture and stability with reference to the surrounding spatial environment.

That part of the inner ear mechanism concerned with hearing is the snail shell–shaped *cochlea* (see Figure 2-42). The cochlea communicates with the middle ear through two small windows: the *fenestra vestibuli* (*oval window*) (which contains the footplate of the stapes of the ossicular chain) and the *fenestra rotunda* (*round window*) located slightly below the oval window and covered with a thin flexible membrane to allow for expansion as fluid movements occur within the cochlea (see Figure 2-42). The cochlea is a tightly packed fluid-filled cavity. It is divided into three channels that extend its entire length from the basal end to the apex; these channels are visible in mid-sagittal dissection of the cochlea (Figure 2-43). The channels are separated by thin, flexible, membranous walls that allow fluid movement in one channel to influence fluid activity in the others. The upper channel (*scala vestibuli*) allows the cochlea to be influenced by the back-and-forth movements of the stapes' footplate within the fenestra vestibuli (see Figure 2-41). It is widely believed that as the footplate of the stapes

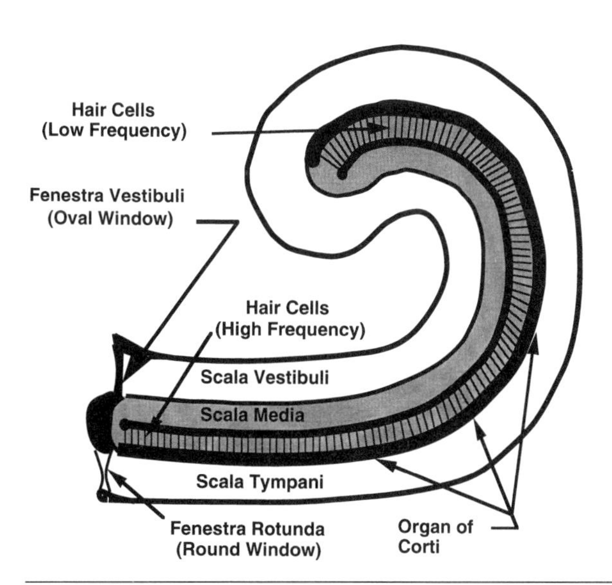

FIGURE 2-43 Cochlea (Longitudinal Section)

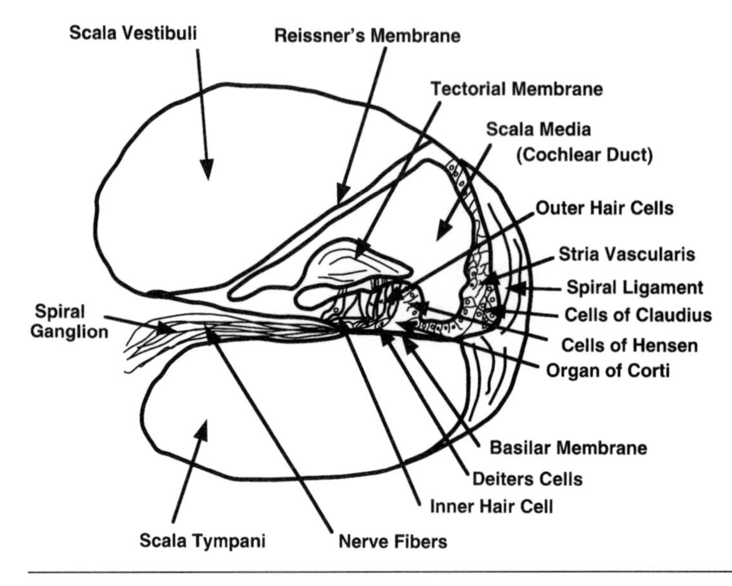

FIGURE 2-44 Cross-Section of Cochlea

moves back and forth within the fenestra vestibuli, the cochlear fluid develops into corresponding wave motions. (This is called the **traveling wave theory.**) These wave motions have an influence on the middle channel (*scala media*), which houses the **organ of Corti,** the sensory end organ of hearing located on the *basilar membrane* (Figure 2-44). It consists of tiny hair cells that are situated in one inner and three to four outer rows

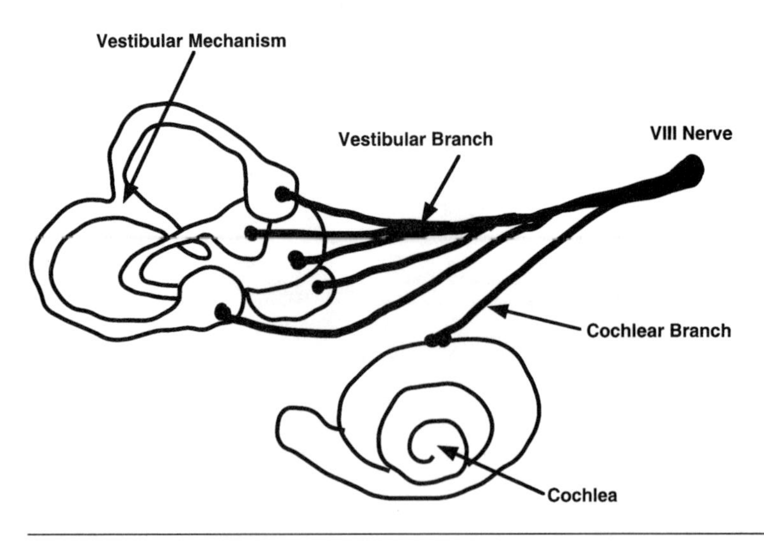

FIGURE 2-45 Vestibular and Cochlear Branches of VIII Cranial Nerve

from the base to the apex of the cochlea. These hair cells are the actual sensory receptor cells for the hearing process. The cilia, small hairlike projections on the tops of the inner and outer hair cells, are embedded in the *tectorial membrane* above.

The hair cells of the organ of Corti contain nerve fibers that group together to form the **cochlear nerve.** The **vestibular nerve** originates from the semicircular canals and joins the cochlear nerve to form the VIIIth cranial (auditory) nerve (Figure 2-45), which sends nerve impulses from the inner ear to the auditory cortex in the brain.

The nerve fibers of the auditory branch of the VIIIth cranial nerve pass along a number of way stations while progressing to the auditory cortex (Figure 2-46). The first way station that marks the beginning of the auditory central nervous system is the *cochlear nucleus.* Other way stations through the brainstem include the *superior olivary complex* in the medulla, the *nucleus of lateral lemniscus* in the pons, the *inferior colliculus* in

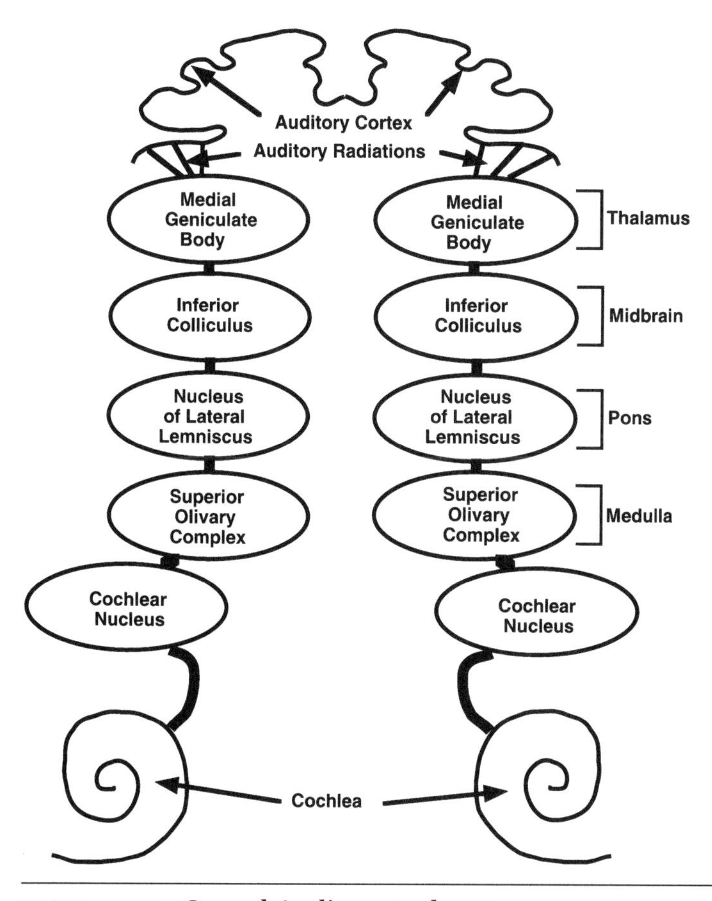

FIGURE 2-46 Central Auditory Pathway

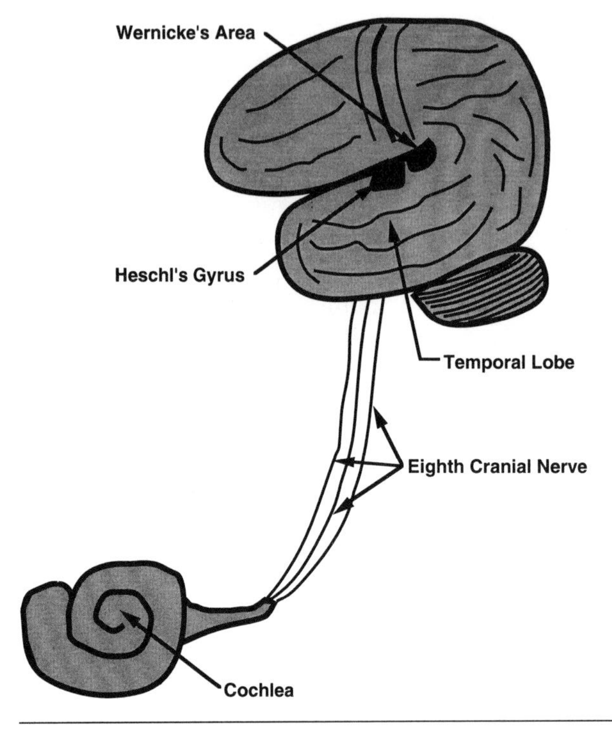

FIGURE 2-47 Cochlea—Temporal Lobe Connections

the midbrain, and the *medial geniculate body* in the thalamus. After this point, the afferent central auditory pathway fans out into multiple small fibers called *auditory radiations,* which connect the medial geniculate body to the **auditory cortex** in the temporal lobe of the brain.

Once the hair cells of the organ of Corti fire, neural signals are sent from the inner ear through the brainstem to the brain via the **VIIIth cranial nerve (acoustic nerve)** (Figure 2-47). The VIIIth cranial nerve connects to an area of the temporal lobe referred to as **Heschl's gyrus,** which has a one-to-one relationship with the hair cells of the organ of Corti; the frequency and intensity information registered at the cochlea is imparted directly to the brain. Once the brain codes this information appropriately, this coded data is shunted to an adjoining temporal lobe area known as **Wernicke's area,** which is connected to those parts of the brain concerned with memory and experience. It is believed that, through a matching process of stored auditory information with the new signals being received from the cochlea, meaning is given to what is being heard.

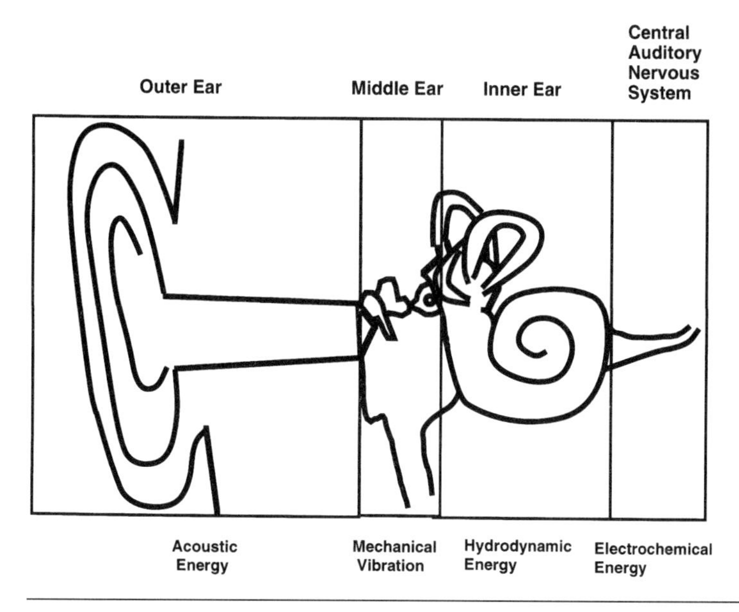

FIGURE 2-48 Cross-Section of Human Ear

Airborne sound undergoes a number of transitions as it is processed through the auditory mechanism (Figure 2-48). It remains airborne as it enters the external auditory meatus, but once the sound strikes the tympanic membrane, its acoustic energy is converted to mechanical vibrations of the membrane itself and to the ossicular chain, which bridges the tympanic membrane and the middle ear cavity located behind it. The ossicular vibrations are converted into fluid wave motions (hydrodynamic energy) in the cochlea, which houses the organ of Corti. The fluid wave motions cause the hair cells of the organ of Corti to be activated. In turn, the organ of Corti converts the fluid motions occurring within the cochlea into neural signals (electrochemical energy), which are transferred to the brain through the VIIIth cranial nerve for final processing.

There are two primary types of theories of hearing. **Place theories** propose that the cochlea serves both as a transducer (converter) of energy (in the organ of Corti) as well as an analyzer of frequency and intensity of sounds. Place theorists believe that for every frequency the ear can perceive, there is a specific place in the cochlea that is sensitive to that frequency. In opposition to place theories are **frequency theories,** which state that pitch and loudness analysis is a central nervous system process, not a cochlear function. Thus, frequency theorists believe that the cochlea serves simply as a transducer of mechanical energy into a neurological code, but the code itself is interpreted by the central nervous system, not by the cochlea. Hearing is a complex process and no theory exists as yet to fully account for all facts associated with audition.

Central Nervous System

The **central nervous system** consists of the **brain** and **spinal cord** (Figure 2-49). The brain is housed within and protected by the skull; it represents the central controlling mechanism for all aspects of human behavior, including speech production and speech perception. The spinal cord, which extends down from the brain through a protective canal within the vertebral (spinal) column, serves to connect major portions of the body to the brain. It is also responsible for many reflexive activities

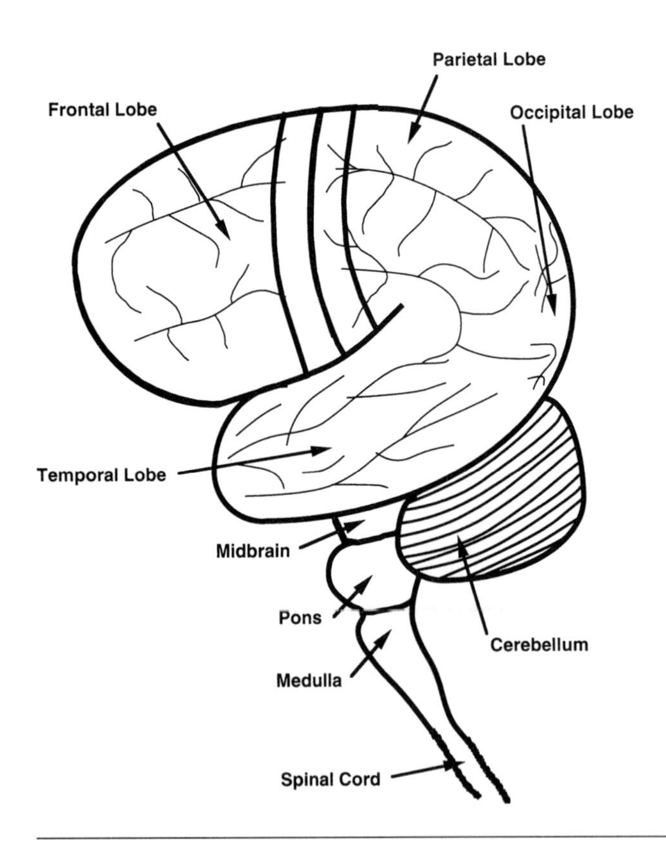

FIGURE 2-49 Brain and Spinal Cord

that do not involve higher thought processes. A number of specific areas of the brain have been identified as being involved in language, speech, and hearing functions. However, the total process by which the brain creates messages, sends them, and receives as well as interprets messages from others, is far from being fully understood.

Hemispheres, Lobes, Landmarks

The brain consists of two *hemispheres* that are connected centrally (Figure 2-50). Each hemisphere consists of a cortical outer layer, which covers various subcortical systems and numerous connecting tracts. The cortical covering (**cortex**) of the hemispheres is composed of millions of cells that function collectively to initiate, control, and interpret behavior. Although

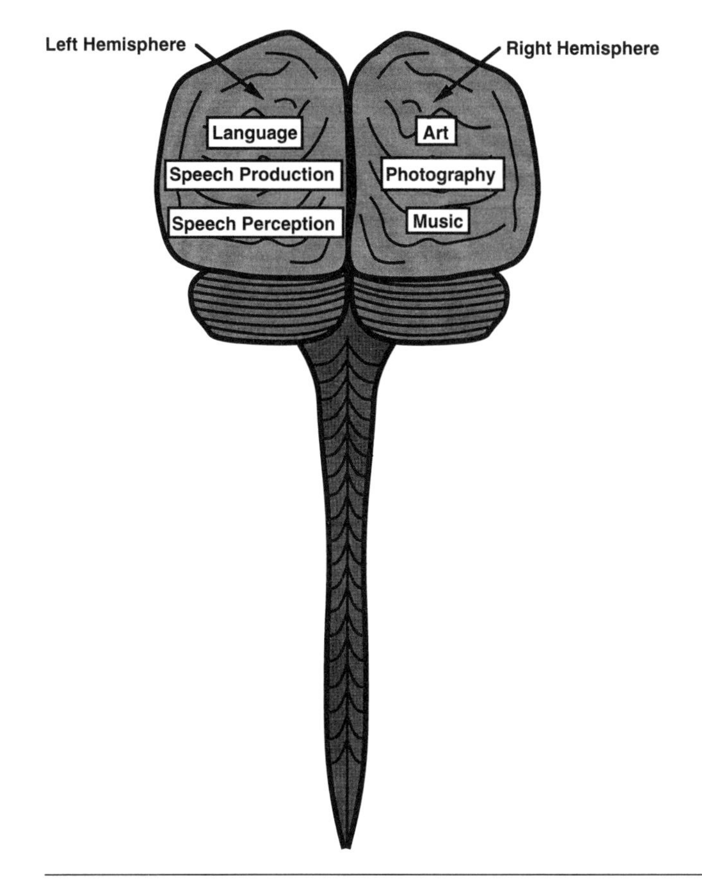

FIGURE 2-50 Hemispheres of the Brain (Posterior View)

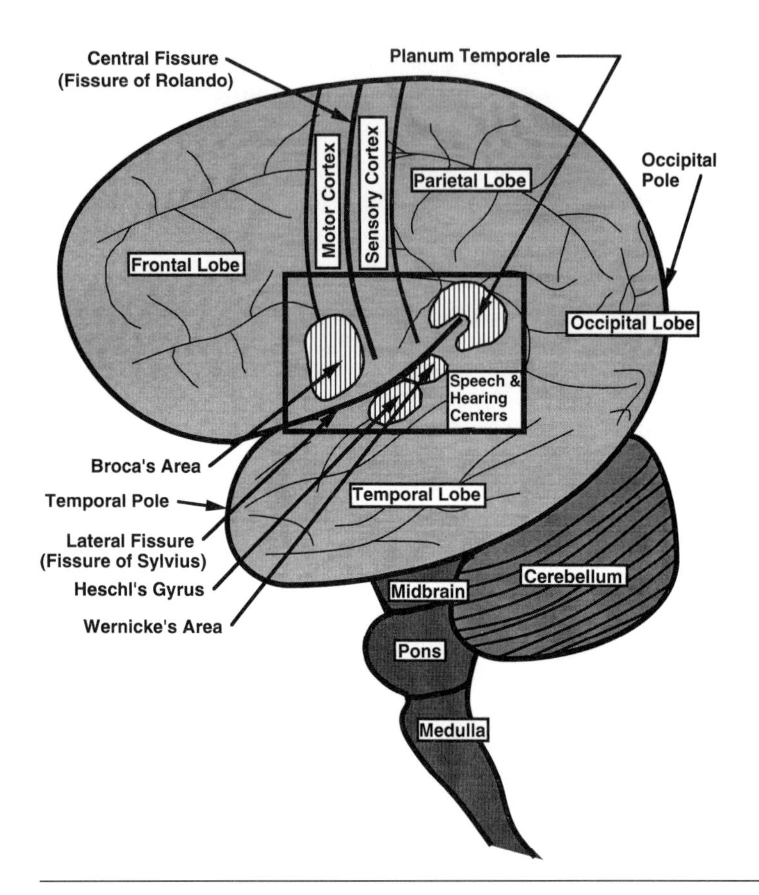

FIGURE 2-51 Lateral View of the Brain (Left Cerebral Hemisphere)

they look similar in appearance and texture, the hemispheres appear to be responsible for different kinds of human activity. The right hemisphere of the brain shows primary responsibility for activities that require appreciation of spatial relationships, such as art, photography, aesthetics, music, and so on. The left hemisphere of the brain appears to relate more to analytical functions, such as mathematics, problem solving, language, speech production, and speech perception. Both hemispheres work together to produce the holistic neurological behaviors that humans can perform routinely. The surface of each of the two cerebral hemispheres is marked by convolutions (raised areas) and sulci or fissures (depressions) (Figure 2-51). Fissures represent the deeper depressions; the two major fissures are the **lateral fissure (fissure of Sylvius),** and the **central fissure (fissure of Rolando).** These two fissures in particular serve as boundary markers for the different lobes of each hemisphere.

Each hemisphere of the brain is divided into five lobes (see Figure 2-51). The **frontal lobe** extends from the *frontal pole* (the most anterior part of the brain) back to the central fissure and downward to the lateral fissure. The frontal lobe is sometimes referred to as the "human" lobe because it appears to be responsible for higher biological thought processing (problem solving), of which animals other than humans are not capable. The frontal lobe, aside from playing a major role in thinking, memory, and language, is also critical to voluntary motor activity. A vertical strip of cortical tissue lying directly in front of the central fissure is the **motor cortex** (*motor strip*) of the frontal lobe and is responsible for initiation of all voluntary motor activity that the body is capable of performing (see Figure 2-51).

The cortical tissue making up the motor strip of the frontal lobe does not provide proportional representation of the human body as we are accustomed to viewing it. More cortical tissue is relegated to the motor control of the the head, neck, arms, and hands than to the total remaining body parts combined. This distorted "brain" picture of the body is called the **homunculus** (Figure 2-52). The bottom of the motor strip,

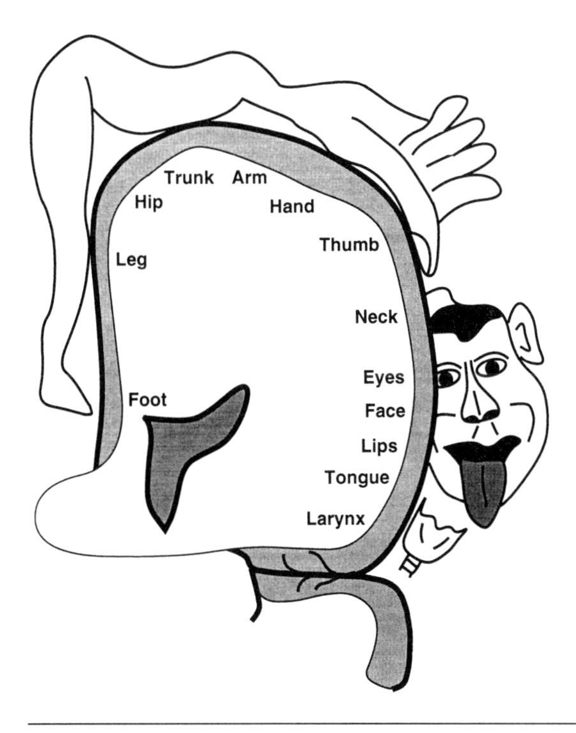

FIGURE 2-52 Homunculus

which is responsible for head and neck mobility, is critical to the motor aspects of phonation and articulation. Activation of the intrinsic laryngeal muscles for phonation, or the tongue and lips for articulation, would begin with cellular activity within this specified location. This lower frontal motor strip region is called **Broca's area** (see Figure 2-51).

The **parietal lobe** lies behind the central fissure and extends back to the **occipital lobe,** which comprises the most posterior portion of each cerebral hemisphere. There is no clear demarcation between the parietal and occipital lobes. The parietal lobe extends downward to the **temporal lobe** from which it also does not have a clear separation (see Figure 2-51). The parietal lobe is often referred to as the "sensory lobe." A strip of *sensory cortex* lying directly in back of the central fissure presents a mirror image of the motor strip of the frontal lobe, which lies directly in front of the central fissure. The sensory strip is concerned with the interpretation of "sensations" from incoming stimuli registered on the skin and other body parts (see Figure 2-51).

The homunculus can be drawn over the sensory strip of the parietal lobe just as it is for the motor strip of the frontal lobe. It seems that those body parts over which we have most motor control are also the ones that show the greatest sensitivity to environmental stimulation. The lip region, for example, will demonstrate much greater sensitivity to touch than will a proportional area located somewhere else within the facial region. Similarly, the index finger will have more tactile precision than will the palm of the hand, the wrist, or the elbow. Other less clearly defined areas of the parietal cortex are related to the learned activities of reading and writing. The scanning of words on a page or the eye–hand coordination required during writing are functions that are related to parietal lobe activity. Overall bodily awareness and body orientation in space are also governed by the parietal lobe. Knowledge of where the head is in relation to the feet, and memory of various body postures are routine behavioral activities requiring parietal lobe monitoring and control.

The occipital lobe is located at the most posterior aspect of the brain (see Figure 2-51). It is bordered by the parietal and temporal lobes and extends backward to the *occipital pole* (the most posterior part of the brain). The occipital lobe is dedicated solely to the sense of vision. The ability to recognize shapes, observe and identify colors, and lock on to visual images entering the visual field are all visuosensory activities under the supervision of the occipital lobe. Other visual functions, such as stereoscopic vision (ability to see depth) and coordinated eye movement, are also under occipital lobe control.

The temporal lobe lies underneath the lateral fissure, extending forward from the occipital lobe to the *temporal pole* (the most anterior part of the brain lying underneath the frontal pole) (see Figure 2-51). The temporal lobe is involved in the senses of smell and hearing. The olfactory bulbs from the nose enter directly into the temporal lobe, as do the tracts from the organ of Corti in the cochlea. As was mentioned, two important cortical regions for hearing, Heschl's gyrus and Wernicke's area, are located within the temporal lobe of the cortex. The temporal lobe also appears to be related to the emotions of fear and bravery. Located within this lobe is a small body of tissue that helps govern an individual's responses to danger and how that individual will function in life-threatening situations.

A fifth lobe, the **insula (island of Reil),** cannot be seen on the surface of a cerebral hemisphere because it is located deep within the lateral fissure. The only way to locate it is to separate the margins of the lateral fissure and look inside. The insula appears to be nature's way of expanding the cortical area of the brain without causing expansion of the skull. It appears to be related to the regulation of gastrointestinal activities such as digestion.

In the left cerebral hemisphere is an enlarged convolution of cortex that surrounds the posterior aspects of the lateral fissure. This convolution is known as the **planum temporale;** it curves around the posterior termination point of the lateral fissure (see Figure 2-51). The planum temporale appears to be an important language coordination center. It lies in a region where numerous connections to the other speech and hearing centers located throughout the left cerebral hemisphere are possible. It is connected to Wernicke's area and Heschl's gyrus, which lie just in front of it on the upper surface of the temporal lobe. The planum temporale also has tracts connecting it to the lower motor strip (Broca's area) of the frontal lobe and the larger anterior expanse of the frontal lobe where memory and experiences are stored. The occipital lobe, which is adjacent to the planum temporale, is also connected to it, providing for visual input to this language coordination center, which can be very helpful in the interpretation of auditory messages in difficult listening situations. Similarly, the bottom of the parietal lobe is adjacent to the planum temporale, providing needed language access for the processes of reading and writing, which are governed by portions of the parietal lobe. A similar convolution exists in the right hemisphere, but it does not appear to be as extensive or as active in language processing activities as the one in the left hemisphere.

Subhemispheric Structures

The **cerebellum** is a lobelike structure lying underneath the occipital lobe (see Figure 2-51). It is not classified as a lobe because it connects to the brain at levels below the cerebral hemispheres. The cerebellum serves as the motor-coordinating mechanism for voluntary motor activity and, as a result, is very important in speech production. Once a voluntary act, such as moving the tongue for the production of a speech sound, is implemented in the frontal lobe motor strip, it is the cerebellum that coordinates the tongue movement, giving it the precision needed for correct speech sound production. If the lips and tongue are moving at the same time for the production of two or more speech sounds (coarticulation), the cerebellum is responsible for integrating that action so that the flow of speech continues without interruption. On a larger scale, the cerebellum is involved in coordinating the three phases of speech production: respiration, phonation, and articulation. The cerebellum is also involved in the regulation of correct muscle tone and various body postures. In addition, it governs the sense of balance and spatial orientation through its connection with the semicircular canals of the inner ear mechanism.

The structures that appear to control and organize all central nervous system activities are located between and below the cerebral hemispheres (Figure 2-53). These structures are found along the midline of the brain just below the tract system (**corpus callosum**) that connects the two cerebral hemispheres. The two **thalami** are ovoid masses, one thalamus on either side of the midline, to which all sensory and motor fibers of a cerebral hemisphere connect. Each thalamus is responsible for the control of all objective behavior (including control of voluntary motor activity for the entire body) as well as the regulation of sensory input from all body parts (including the special senses of vision and hearing). Some experts believe that the thalami are small brains within themselves and that the cerebral hemispheres with their cortical coverings act as information storage units to be accessed as necessary.

The **hypothalamus** is a singular, less well-defined structure lying on the midline of the brain just below and in front of the thalamus (see Figure 2-53). The hypothalamus appears to be in charge of animalistic functions such as eating, consummatory behavior, sleeping, sexual behavior, and temperature regulation. This tiny structure controls these basic activities through control of the pituitary gland, which is responsible for distribution of hormones throughout the body.

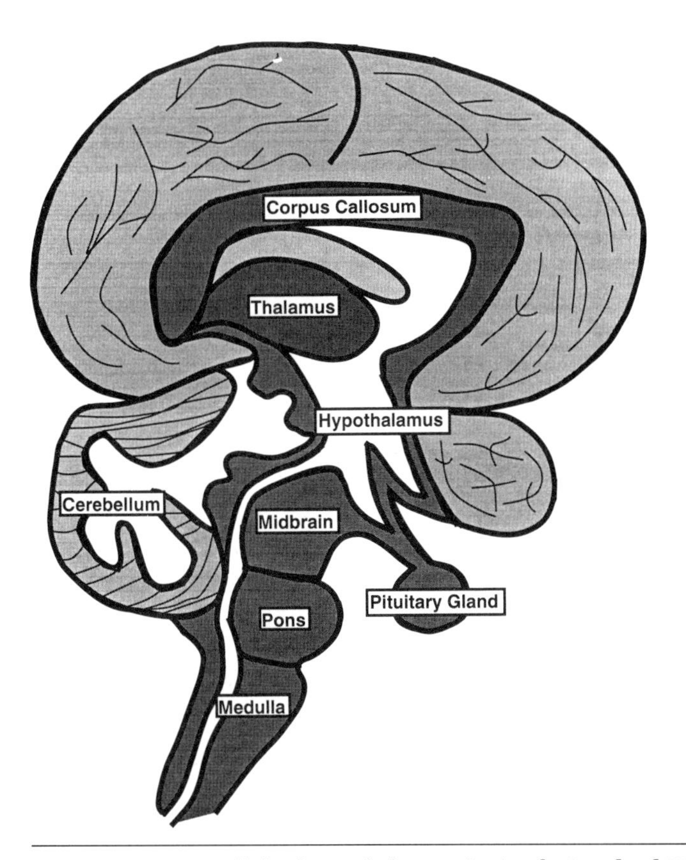

FIGURE 2-53 Medial View of the Brain (Left Cerebral Hemisphere)

The **medulla** is a section of the brain located at the low end of the brain where it blends into the spinal cord (see Figures 2-51 and 2-53). It contains life-support centers concerned with respiration and circulation. This brainstem structure is important for speech production because of the need for an energy source (air supply) to produce speech. The respiratory centers within the medulla can be temporally controlled by higher brain mechanisms during the act of exhalation for speech production. During speech production, the exhalatory cycle of breathing is extended according to phrasing necessities, while inhalation is inserted into juncture points and pauses. During normal (vegetative) breathing, the inhalation and exhalation cycles are similar in duration and automatically regulated at the medullary level.

Peripheral Nervous System

The peripheral nervous system consists of the **cranial** and **spinal nerves.** Its purpose is to provide connections for the head and body to the central nervous system. The cranial nerves are 12 pairs of highly specialized nerves that are designed primarily to connect the major motor and sensory elements of the head and neck to the central nervous system. The spinal nerves are 31 pairs of nerves that course out from the spinal cord to all parts of the body. They convey motor signals to the body from the central nervous system and gather sensory input from the body for central nervous system processing.

Cranial Nerves

The cranial nerves can be described as individual units with specific motor and sensory functions. Not all of the cranial nerves are related to speech production and speech perception, but those that are related are critical to the process. All of the cranial nerves are listed with brief functional descriptions in Figure 2-54. Those cranial nerves involved in speech production and speech perception processes are the *trigeminal (V), facial (VII), acoustic (VIII), glossopharyngeal (IX), vagus (X),* and *hypoglossal (XII).*

The trigeminal (V) nerve is both a motor and a sensory nerve. It provides motor connections for the muscles of mastication that are responsible for mandibular (lower jaw) movement, which is important in the articulatory process. The mandible also serves as a platform for the tongue (which is the major articulator). Fine motor movements in one structure will have an effect on the activities of the other. The trigeminal nerve is a sensory nerve to the face, sinuses, and teeth.

The facial (VII) nerve is responsible for motor movements of the facial muscles. Muscles such as the orbicularis oris (which makes up the sphincter ring of the lips) are very important for articulation. This nerve provides sensory input from the anterior two-thirds of the tongue as well as from the velum for tactile and proprioceptive feedback monitoring purposes during the articulatory process.

The acoustic (VIII) nerve is a sensory nerve that is totally responsible for delivering auditory information from the inner ear (cochlea) to the temporal lobe area of the brain (Heschl's gyrus). If this nerve is damaged or severed, hearing will be lost in the involved ear.

The glossopharyngeal (IX) nerve provides motor function to the pharyngeal musculature, which helps shape the posterior aspects of the vocal tract during articulation. The walls of the pharynx can be constricted to help form a seal with the soft palate to close off the nasal cavity. The back wall of the pharynx can be brought forward to come into

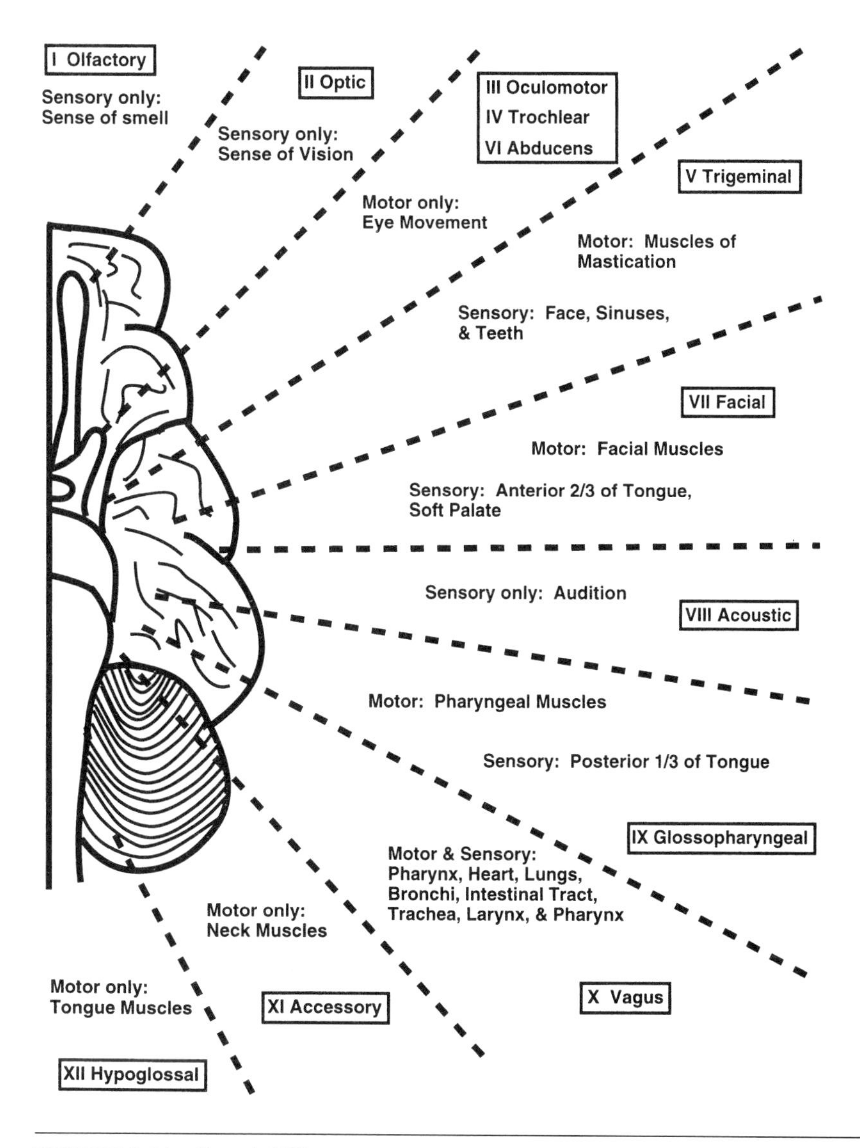

FIGURE 2-54 Cranial Nerves

proximity with the back of the tongue for the production of various speech sounds. The glossopharyngeal nerve provides sensory feedback from the posterior one-third of the tongue. This nerve, combined with the facial nerve, helps provide tactile sensory input to the entire lingual surface. However, lingual sensory feedback from the glossopharyngeal nerve does not appear to be as acute as that from the facial nerve. There is general agreement that it is easier to feel the tongue tip touch the teeth or alveolar ridge than it is to feel the back of the tongue come into contact with the velum or pharynx.

The vagus (X) nerve seems to travel everywhere throughout the body. It is an exception to the concept that cranial nerves serve only to connect the head and neck to the central nervous system. The vagus nerve serves in a motor and sensory capacity for the intestinal tract, heart, lungs, bronchi, trachea, larynx, and pharynx. It also serves as a tactile sensor for the external ear. In addition, it is critical to laryngeal operation for phonatory purposes. All of the intrinsic muscles of the larynx are governed by the vagus nerve. In a sensory capacity it allows the larynx to react to foreign matter, causing a coughing action for ejection purposes. If damage occurs to any or all parts of the vagus nerve, various kinds of paralysis of one or both of the vocal folds will occur.

The hypoglossal (XII) nerve is a motor nerve solely responsible for tongue mobility. All of the intrinsic muscles of the tongue are innervated by this nerve and rely on it for the smooth, intricate tongue positioning needed for speech articulation. Damage to this nerve will cause various types of paralysis to one or both sides of the tongue.

It can be argued that the optic (II) nerve should be considered important to the communication process. Individuals who are severely or profoundly hearing impaired rely almost totally on vision to replace their hearing. They communicate through sign language and speechreading, and will use all types of visual cues to receive and understand a message. Even normal-hearing individuals take advantage of visual cues in the form of pictures, signs, and symbols to augment what they have heard, especially in difficult listening situations.

Spinal Nerves

The 31 pairs of spinal nerves emerge from the spinal cord and pass between the individual vertebra of the vertebral column into the body proper (Figure 2-55). These nerves are divided into groupings with the same names as those used to differentiate vertebrae in the vertebral column. There are 8 pairs of cervical, 12 pairs of thoracic, 5 pairs of lumbar, 5 pairs of sacral, and 1 pair of coccygeal spinal nerves. Each spinal nerve is connected to the spinal cord by 2 pairs of roots: anterior and posterior. The anterior roots are made up of nerve fibers carrying nerve impulses away from the spinal cord. These motor fibers course throughout the body, providing motoric stimulation to the appropriate musculature. The posterior roots consist of nerve fibers carrying nerve impulses to the cen-

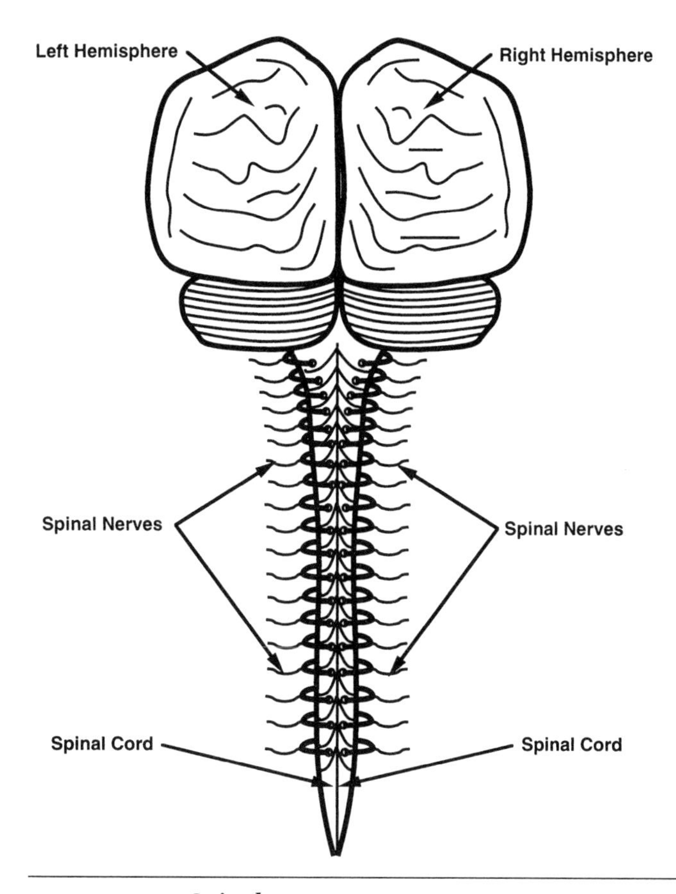

FIGURE 2-55 Spinal Nerves

tral nervous system. These sensory fibers approach the spinal cord from all aspects of the body, providing it with appropriate sensory information from all body regions.

Autonomic Nervous System

The **autonomic nervous system** is an independent entity. It consists of chains of nerve cells lying on either side of the spinal cord (Figure 2-56). Although it is connected to the spinal cord and will communicate with it when time permits, the autonomic nervous system can act independently of the central nervous system when there is insufficient time for conscious decision making to take place. It is divided into two parts: sympathetic and parasympathetic. The **sympathetic** part helps prepare the body for emer-

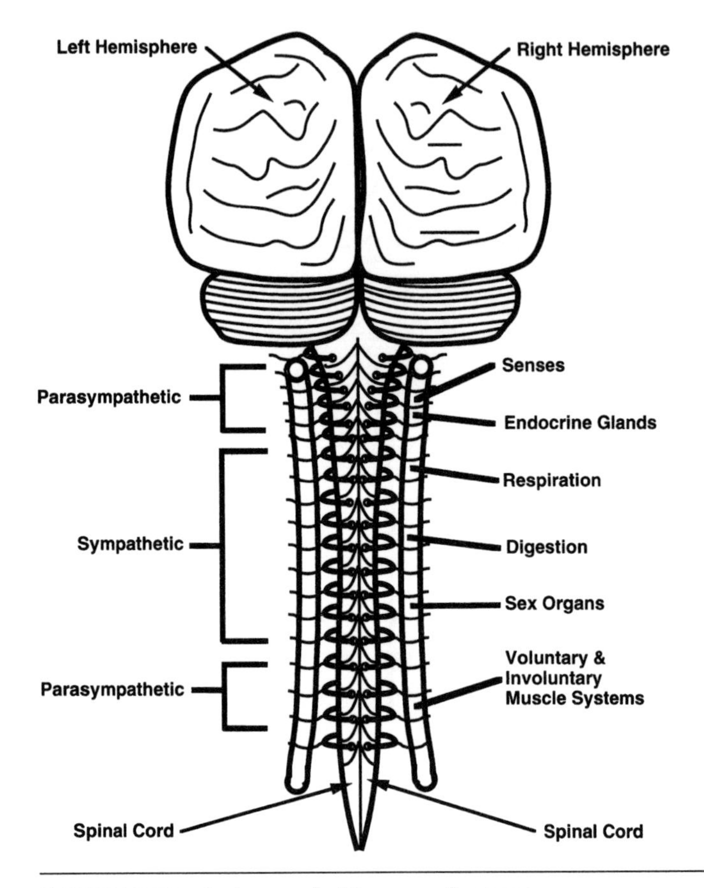

FIGURE 2-56 Autonomic Nervous System

gency situations. If confronted with danger, the body can be activated quickly in terms of sympathetic alterations in heart rate, muscle strength and quickness, dilation of the pupils, voice, articulation, hearing, and so on. Once the danger has passed, the various alterations in these systems can be returned to a more normal state (**homeostasis**) by the **parasympathetic** part of the autonomic nervous system. Parasympathetic return to a more normal state will occur over time and is not as immediate as the sympathetic action needed to survive emergency situations.

The autonomic nervous system plays a role in reflecting an individual's emotional states through the speech and hearing mechanisms. If a person becomes excited, there will be changes in breathing patterns (respiratory cycles), voice patterns (frequency and intensity), and articulatory patterns (intelligibility) commensurate with the stress or urgency being felt. Similarly, when the individual returns to more normal conditions, breathing rates, voice shifts, and articulatory precision will be altered accordingly. The voice, in particular, has been referred to as the major conveyer of emotions and seems to be specifically affected by autonomic nervous system activities.

Hearing can also be influenced by the autonomic nervous system. When in excited states of emotion, it is sometimes very difficult for individuals to listen or take appropriate direction because of their focus on the survival task at hand. However, when more relaxed, these same individuals are often better able to analyze what they are hearing and give more appropriate meaning to incoming signals. At other times, individuals appear to show more acute hearing abilities in emergency situations. They seem to be able to process more information faster and more efficiently than when more relaxed. The subjective emotional states that people assume on a daily basis can have a definite influence on all aspects of speech production as well as the processing of incoming speech signals.

Summary

The study of the anatomy and physiology of the speech and hearing mechanisms provides an appreciation of the body tissues involved in the communication process. With the exception of the hands and legs, all other parts of the human system are included to some degree in the acts of speech production and speech perception.

Speech

Speech production requires synchronization of the processes of respiration, phonation, and articulation. Respiration provides the exhaled air supply needed for speech production; phonation is concerned with vibrating the exhaled air for sound production purposes; and articulation provides the shaping of the sound produced into the individual speech sounds of a language.

Hearing

The hearing mechanism is responsible for the initial learning of speech from others, the monitoring of one's own speech, and the reception of the acoustic speech signal from other speakers for communication purposes. Anatomically, the hearing mechanism consists of the outer ear, the middle ear, and the inner ear.

As airborne sound travels through the external auditory meatus and strikes the tympanic membrane, the sound is transformed into mechanical vibrations of the membrane itself. The ossicles transfer the mechanical vibrations of the tympanic membrane to the inner ear mechanism where the sensory end organ of hearing, the organ of Corti, is located. The organ of Corti consists of tiny hair cells that respond to fluid movement within the cochlea. The fluid movement is set up by the back-and-forth movement of the ossicles, which communicate with the cochlea through the oval window. Once the hair cells within the cochlea are activated, they send neural signals to certain parts of the brain. By the time airborne sound is processed by the hearing mechanism, it has gone from being an acoustic signal to mechanical vibrations, from mechanical vibrations to fluid movement, and from fluid movement to neural signals, which are transferred to the brain for final processing.

Central Nervous System

The central nervous system consists of the brain and spinal cord. The brain is the central controlling mechanism for all aspects of human behavior, including speech production and speech perception. The brain consists of two cerebral hemispheres. The right hemisphere is responsible for activities involving spatial relationships, while the left hemisphere is responsible for analytical tasks (problem solving, language, speech production, and speech perception).

The structures that appear to control and organize all central nervous system functions are the thalami and the hypothalamus. The medulla, which houses the centers for respiration and circulation, is located at the lower end of the brain, where it blends into the spinal cord.

Peripheral Nervous System

The peripheral nervous system consists of cranial and spinal nerves. Its purpose is to connect the head and body to the central nervous system. The cranial nerves are 12 pairs of highly specialized nerves designed primarily to connect the major motor and sensory aspects of the head and neck to the central nervous system. Those cranial nerves involved in speech production and speech perception are the trigeminal (V), facial (VII), acoustic (VIII), glossopharyngeal (IX), vagus (X), and hypoglossal (XII). The 31 pairs of spinal nerves convey motor signals to the body from the central nervous system and collect sensory input from the body for central nervous system interpretation.

Autonomic Nervous System

The autonomic nervous system can act independently of the central nervous system during emergency situations. The sympathetic part of this system helps prepare the body for emergencies, and once the danger is passed, the parasympathetic part of the system helps the body to return to a more normal state. The autonomic nervous system plays a role in reflecting an individual's emotional condition through the speech and hearing mechanisms. Respiratory, phonatory, and articulatory patterns are affected by the amount of stress or urgency that an individual is feeling at any given point in time. The voice in particular is a major indicator of emotional states.

Study Questions

1. What skeletal structures are involved in the respiratory process?
2. In primarily what region of the body are the muscles that are involved in the inhalation phase of respiration? the exhalation phase of respiration?
3. What are the three unpaired and three paired cartilages that compose the larynx?
4. List the extrinsic larygeal elevator and laryngeal depressor muscles of the larynx.
5. The intrinsic laryngeal muscles include those that abduct, adduct, tense, and relax the vocal folds. Name these muscles.
6. The articulators, those structures that lie above the larynx within the cavities of the pharynx, oral cavity, and nasal cavity, change the shapes of these cavities and thereby convert sounds into speech sounds. List the articulators and describe their specific role in the articulatory process.
7. The peripheral auditory mechanism can be divided anatomically into three parts:
   a. ____________________
   b. ____________________
   c. ____________________
8. The auditory mechanism can be divided functionally into three main parts:
   a. ____________________
   b. ____________________
   c. ____________________

9. The inner ear contains two sensory end organs; the one for hearing inside the cochlea is the ___________________; the one(s) for body balance/equilibrium is (are) the ______________________.
10. Describe the path of the auditory signal from the outer ear to the auditory cortex. Include a delineation of the various way stations in the central auditory pathway.
11. Identify and describe the two most popular theories of hearing and delineate their primary differences.
12. The central nervous system consists of two structures:
    a. ____________________
    b. ____________________
13. Each hemisphere of the brain is divided into five lobes:
    a. ____________________
    b. ____________________
    c. ____________________
    d. ____________________
    e. ____________________
14. What is the autonomic nervous system, and what is its role in speech production and speech perception?

Suggested Readings

Abbs, J.H. (1996). Mechanisms of speech motor execution and control. In Lass, N.J. (Ed.), *Principles of Experimental Phonetics.* St. Louis: Mosby, pp. 93-111.

Fucci, D., & Petrosino, L. (1984). The practical applications of neuroanatomy for the speech-language pathologist. In Lass, N.J. (Ed.), *Speech and Language: Advances in Basic Research and Practice, Vol. 11.* New York: Academic Press, pp. 249–313.

Greenberg, S. (1996). Auditory processing of speech. In Lass, N.J. (Ed.) *Principles of Experimental Phonetics.* St. Louis: Mosby, pp. 362–407.

Orlikoff, R.F., & Kahane, J.C. (1996). Structure and function of the larynx. In Lass, N.J. (Ed.), *Principles of Experimental Phonetics.* St. Louis: Mosby, pp. 112–181.

Seikel, J.A., King, D.W., & Drumright, D.G. (1997). *Anatomy and Physiology for Speech, Language, and Hearing.* San Diego: Singular Publishing Group.

Warren, D.W. (1996). Regulation of speech aerodynamics. In Lass, N.J. (Ed.), *Principles of Experimental Phonetics.* St. Louis: Mosby, pp. 46–92.

Zemlin, W.R. (1998). *Speech and Hearing Science: Anatomy and Physiology* (Fourth Edition). Boston: Allyn and Bacon.

3

Basic Acoustics

An understanding of the physics of sound is essential to an understanding of the speech signal, which is a complex sound. This chapter addresses basic concepts associated with sound. Its purpose is to help the reader gain insight into basic acoustics, which can then be applied to an understanding of the acoustics of speech production.

Sound

Sound can be defined as a condition of disturbance of particles in a medium. Air is the medium most often used for human speech production. If a portion of air could be scrutinized microscopically, it would be found to consist of billions of air particles, called **molecules** (Figure 3-1). A further discovery would be that these molecules are nearly equally

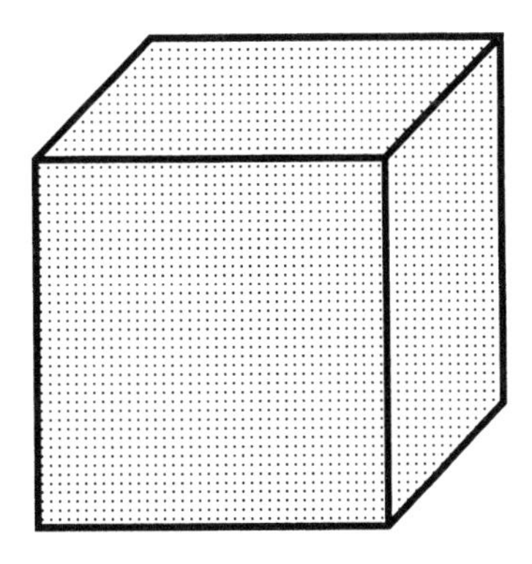

FIGURE 3-1 Air Molecules

spaced with respect to one another. The properties common to the medium of air and other media used for the transmission of sound waves are mass, elasticity, and inertia.

Mass is any form of matter (solid, liquid, gas) capable of vibratory motion. The particles in a medium like air consist of mass. If a medium has **elasticity,** it is able to resist permanent distortion to its original shape or the distribution of its particles. Thus, it possesses the property of "springiness." Since air is not observable, it is difficult to think of it in terms of having a shape that can be distorted. A visual aid useful for an understanding of these concepts is the spring-mass model, as shown in Figure 3-2. The molecules (mass) in air behave as if they had springs attached to them (as illustrated in Figures 3-3A and 3-3B), which allows them to be moved from and returned to their original rest position. However, because of inertia, they do not stop. **Inertia** is a property common to all matter: A body in motion will remain in motion, while a body at rest will remain at rest (unless acted upon by an external force). Since the molecules are in motion as they move toward their rest position, they will not stop at this position, but continue to move beyond it. This is illustrated in Figure 3-2 using the spring-mass model.

In addition to a transmitting medium, the other elements necessary for the production of sound include an energy source and a vibrating body. An energy source is used to activate a vibrator of some kind; the

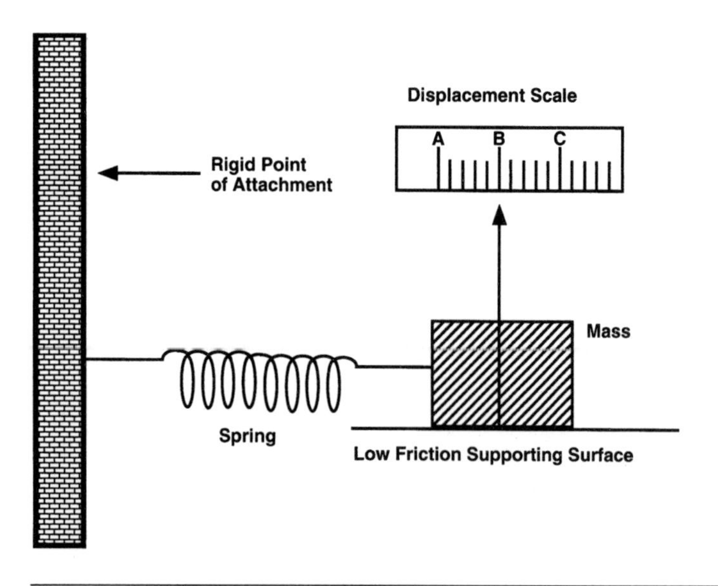

FIGURE 3-2 Spring-Mass Model

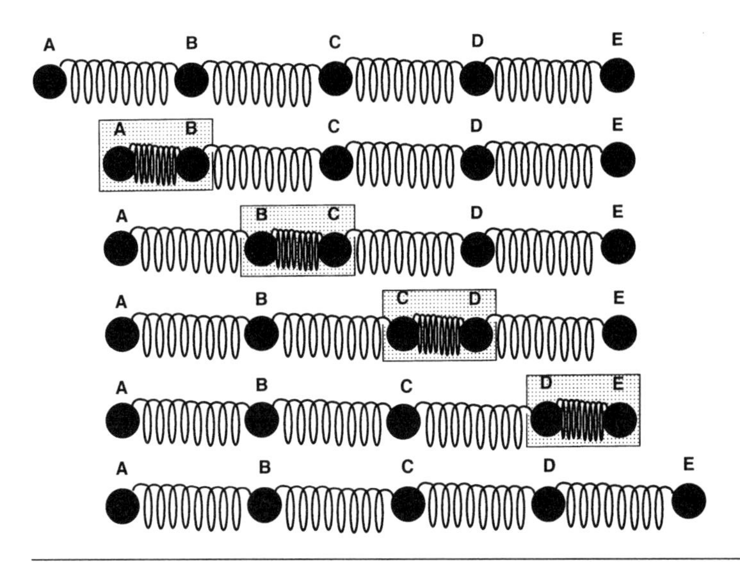

FIGURE 3-3A Air Molecules: Compression

energy source required often depends on the vibrator itself. A vibrator such as a tuning fork needs to be struck against a hard surface in order for it to be activated. Drum heads need to be hit with a stick or mallet to cause disturbances in the medium. If air is forced between tightly

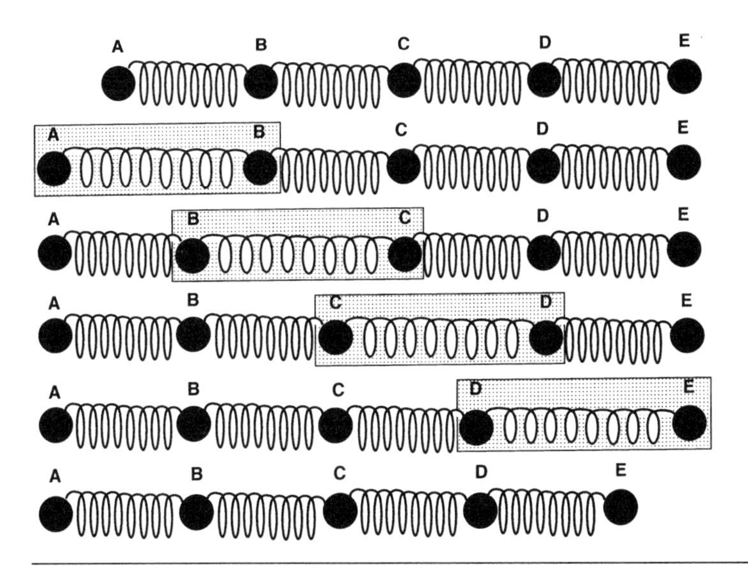

FIGURE 3-3B Air Molecules: Rarefaction

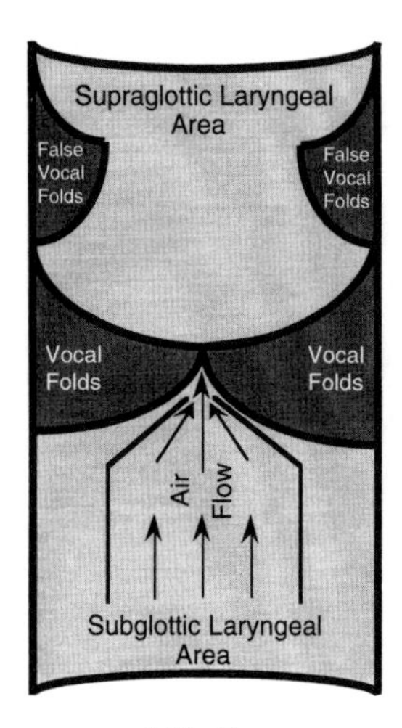

FIGURE 3-4 Air Flow During Phonation

constricted lips, a buzzing sound can be made, which is used as a sound source for trumpet and tuba players. Air is also the primary source of energy for speech production. A steady stream of air flowing from the lungs is required for the vocal folds of the larynx (the vibrator for all voiced speech sounds) to open and close rapidly enough to produce frequencies within the range of human hearing (Figure 3-4). The number of openings and closings of the folds that occur in one second determines the perceived pitch of the voice.

The human sound-producing mechanism has the necessary ingredients to produce sound (an energy source, vibrator, and transmitting medium) as well as an additional component that enables it not only to produce sound, but also to convert sound disturbance into speech sounds (vowels, consonants, diphthongs) that are reflective of a particular language. This speech sound component is called the **vocal tract** (Figure 3-5), which extends from the vocal folds of the larynx through the pharynx, oral cavity, nasal cavity, and lips. (The vocal tract is discussed in detail in Chapter 5.)

Sinusoidal Motion

Describing sound in such a way as to visualize it is not a straightforward process because of the abstract nature of the concept of sound. One way

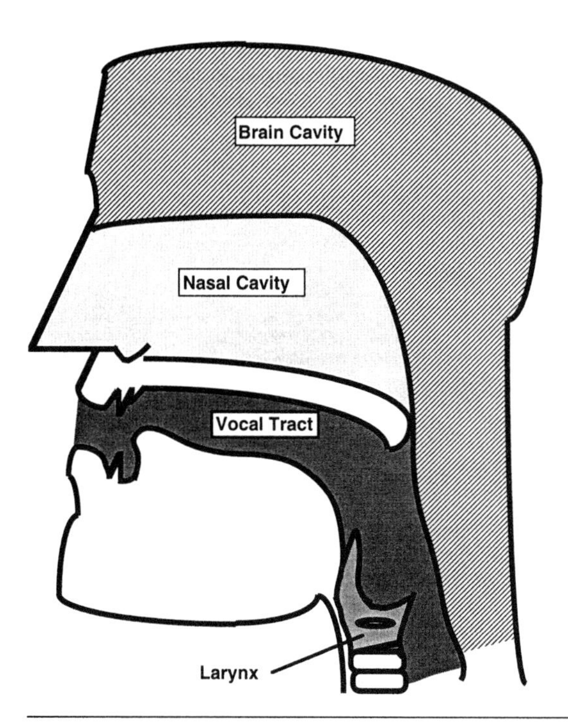

FIGURE 3-5 Vocal Tract

is by discussing the simplest kind of sound wave motion that can occur in a medium. This simple wave motion is called **sinusoidal motion.** Sinusoidal motion, also called **simple harmonic motion,** is a disturbance in a medium that occurs when devices such as tuning forks and clock pendulums are activated. Figure 3-6 illustrates sinsusoidal motion as it is being traced from the movements of a clock pendulum. If a sheet of paper could be pulled underneath the back-and-forth movements (oscillations) of a swinging pendulum with a pen attached to the bottom of it, the picture of a sine wave would emerge on the paper. The pendulum would begin its movement from a point of rest, move in one direction to a point of maximum displacement, return to its point of rest, go through its point of rest to maximum displacement in the opposite direction, and then return again to its rest position. The result is a sine wave tracing.

This **sine wave** or **sine curve** (named because it is derived from certain trigonometric functions of right angles) is used to represent a sound wave in general; however, it is also used to characterize the movement of particles in a medium as a result of the disturbance that created the sound. The features of the particles represented include *particle pressure, particle*

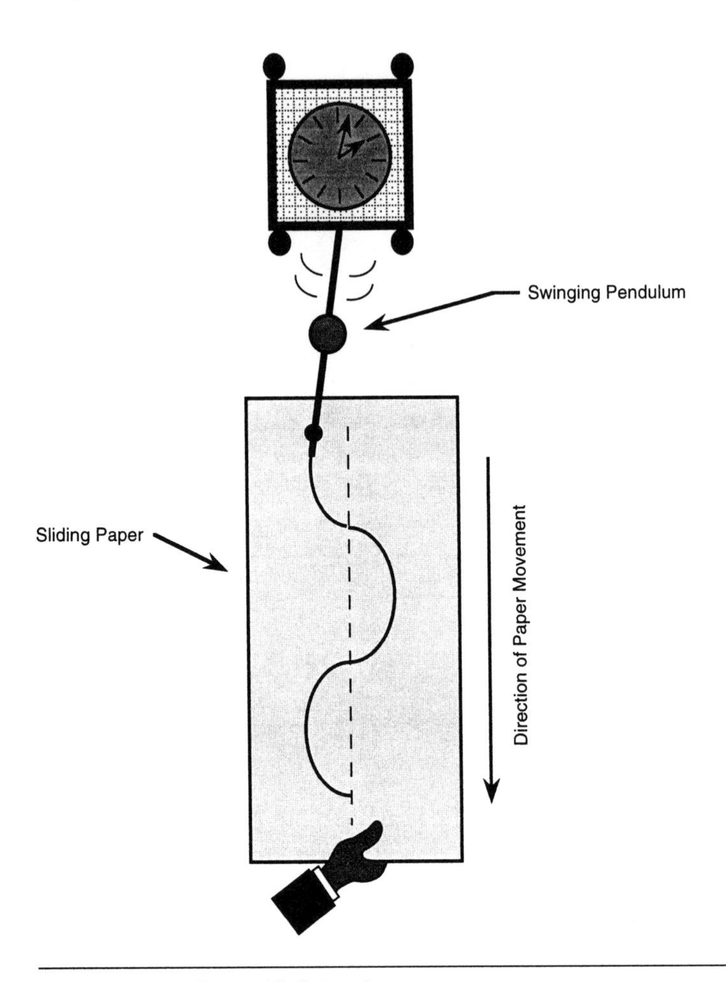

FIGURE 3-6 Sinusoidal Motion

velocity, and *particle displacement.* Figure 3-7 shows these properties of particles in a medium and their relationship to each other. As the figure illustrates, maximum particle pressure and maximum particle displacement are identical, and both are inversely related to maximum particle velocity.

The sound that is generated from vibrators that produce sinusoidal movement is often designated as a **pure tone,** a sound that has all of its energy located at one frequency. Pure tones are rarely heard in everyday situations; most of the sounds that we routinely hear in our environment are complex in that their energy is concentrated at more than a single frequency.

When sinusoidal wave motion disturbs the particles of the medium, they react in a predictable way. As the pendulum or tuning fork tine begins

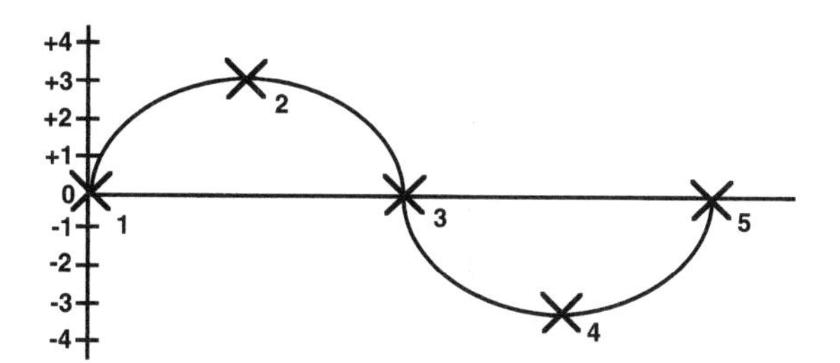

1, 3, 5 = zero displacement
1, 3, 5 = zero pressure
1, 3, 5 = maximum velocity
2, 4 = Maximum Displacement
2, 4 = maximum pressure
2, 4 = minimum velocity

FIGURE 3-7 Particle Displacement, Pressure, and Velocity

to move from rest to maximum displacement in one direction, the particles in the medium are pushed closer toward each other; they are said to be in a state of **compression** (see Figure 3-3A) or **condensation** (see Figure 3-3B). Maximum compression takes place at the point of maximum excursion of the vibrating pendulum or tuning fork tine. As the pendulum or tuning fork tine begins to move in the opposite direction, the particles attempt to return to their original positions (because of elasticity), but they overshoot that position (because of inertia) before coming to rest again; this overshoot, wherein the particles are spread apart more than they normally would be, is called a state of **rarefaction** or *expansion* (Figures 3-3B and 3-8). These condensations and rarefactions are the actual sound disturbances that travel through the medium from the sound source. It should be noted that the particles (molecules) themselves are not moving through the medium. The particles near a person's lips during sound production will move around their points of origin (rest positions), but once the sound disturbance has traveled away from the lips, those particles will return to their rest positions. Thus, the disturbance will have moved away from the lips, but not the individual particles in the medium; they will simply be displaced temporarily from their rest positions.

The sine wave tracing can provide a spatial or a temporal picture of particle disturbances in the medium. As a spatial picture, the sine wave tracing indicates the relative positions of the particles in the medium at a single instant in time. As a temporal picture, it can be used to study the

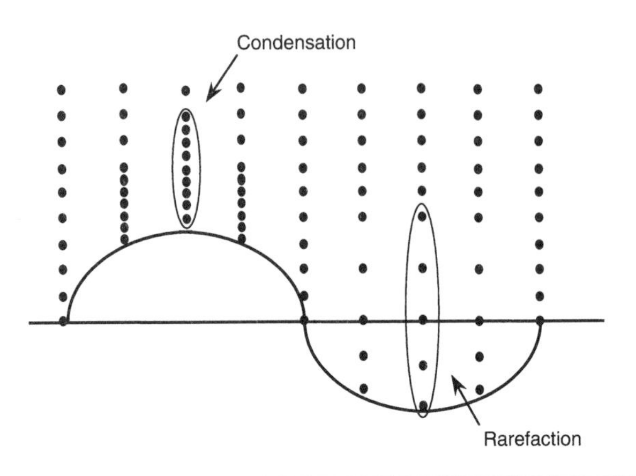

FIGURE 3-8 Sine Wave Disturbance of Medium Particles

movement of a single particle over time as it changes its position around its rest position. Each view of the sine wave tracing has a set of terms associated with it.

Spatial Concepts

Amplitude

Amplitude refers to the maximum displacement of the particles of a medium. It is related perceptually to the magnitude of the sound (volume, loudness). Amplitude indicates the energy (intensity) of a sound; it is usually measured from the baseline (point of rest) to the point of maximum displacement on the wave form (Figure 3-9). This linear measurement is called **peak amplitude** measurement. The distance between the baseline and the point of maximum displacement is related to the movement of the swinging pendulum or tuning fork tine as it moves from rest to maximum excursion in one direction. In other words, amplitude is related to the point of maximum vibration of a particular vibrating object. In the case of the spring-mass model, it represents maximum excursion of the mass from its rest position (see Figure 3-2). In the case of the human vocal folds, the amplitude of the sound being produced is related to the maximum excursion of the vocal folds away from the midline during each cycle of vibration. The farther the folds are spread apart during each cycle, the greater is the resulting amplitude of the sound being produced (Figure 3-10).

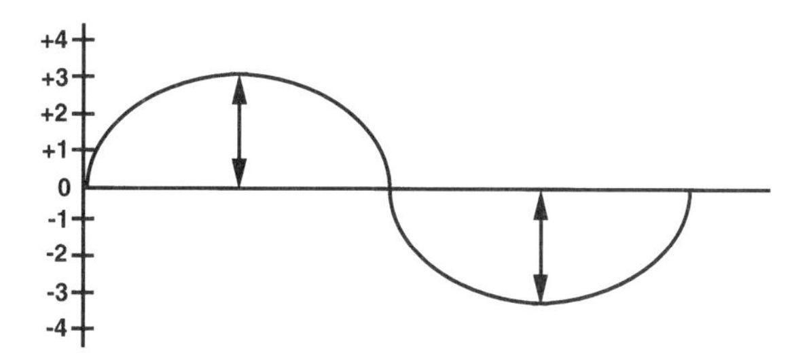

FIGURE 3-9 Amplitude of a Sine Curve

In some instances, amplitude measurements are made on the sine wave tracings from the point of maximum displacement in one direction to the point of maximum displacement in the other direction, instead of from baseline to the point of maximum displacement in one direction. This linear measurement is called **peak-to-peak amplitude** measurement (Figure 3-11). It makes sense that the actual peak-to-peak amplitude reading would be twice that of the peak amplitude reading mentioned earlier

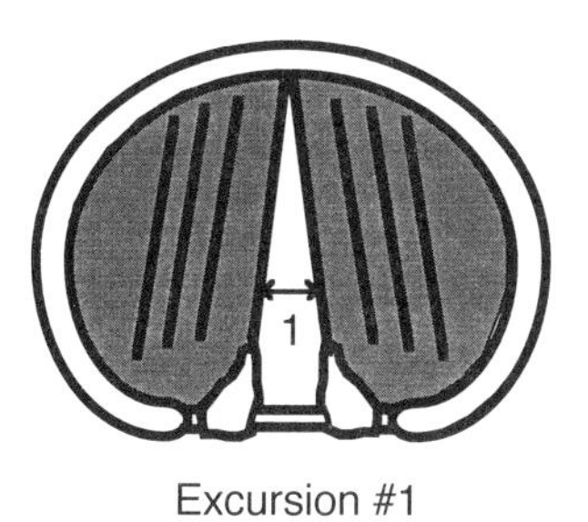

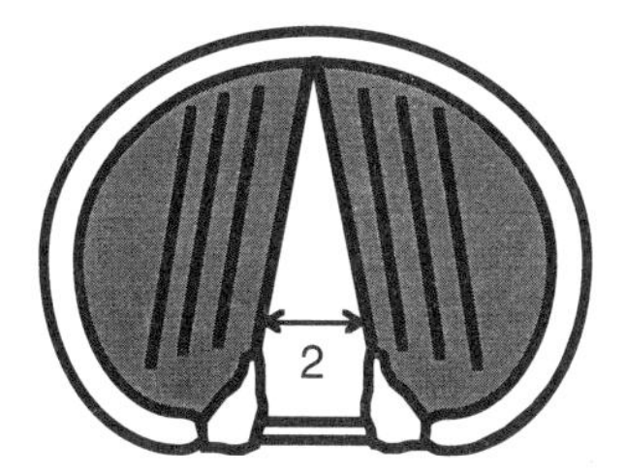

FIGURE 3-10 Vocal Fold Excursions

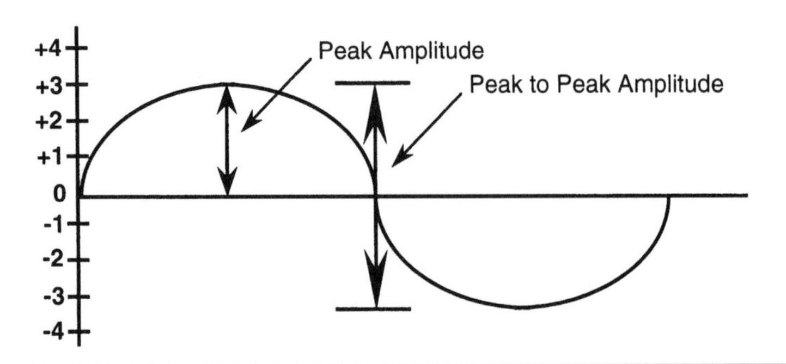

FIGURE 3-11 Peak Amplitude and Peak-to-Peak Amplitude of a Sine Curve

(and shown in Figures 3-9 and 3-11). The amplitude in both instances would be the same, but the method of measurement would be different, leading to different linear readings. Therefore, it is important to indicate whether the amplitude being reported is in terms of peak or peak-to-peak measurements.

Amplitude is related to the measurement of intensity, which can be expressed in terms of sound pressure level or power. The decibel (dB) is the most common unit used to express sound intensity when amplitude is being expressed in terms of sound pressure or power. It is a logarithmic unit used to express ratios between pressures or powers of sounds.

Wavelength

Wavelength (λ), another spatial term, is a linear measurement that refers to the distance a sound wave disturbance can travel during one complete cycle of vibration. More specifically, wavelength can be defined as the dis-

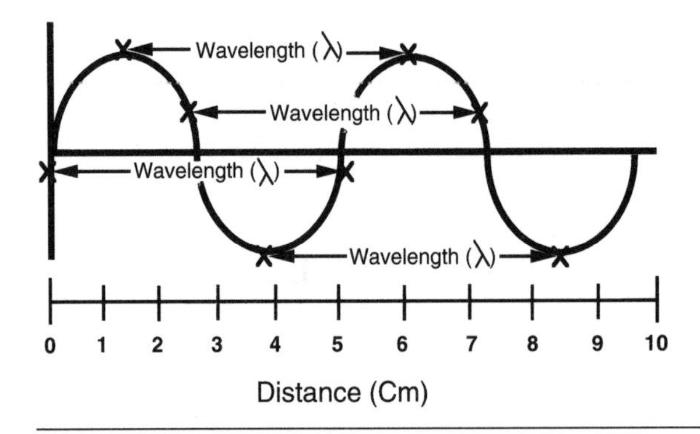

FIGURE 3-12 Wavelength

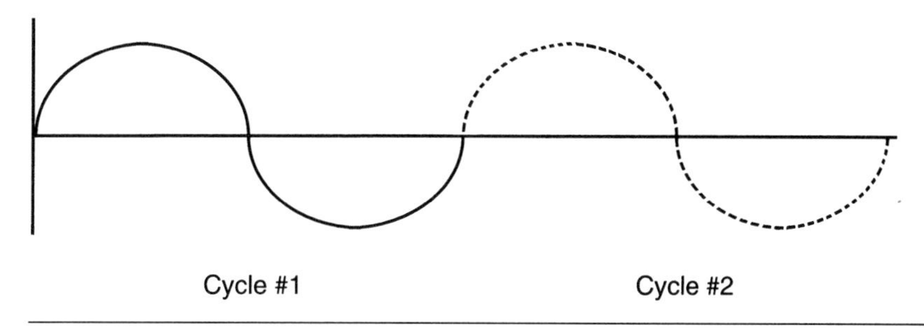

FIGURE 3-13 Cycles of a Sine Curve

tance between points of identical phase in two adjacent cycles of a wave (Figure 3-12). It can be expressed in feet or meters, and, as will be seen later in this chapter, it is inversely related to the frequency of the sound being produced.

Temporal Concepts

Cycle

Cycle is a time concept referring to vibrator movement from rest position to maximum displacement in one direction, to rest, to maximum displacement in the opposite direction, and back to rest again (Figure 3-13).

Period

Period is the time (usually expressed in seconds) that it takes for a vibrator to complete one entire cycle of vibration (Figure 3-14).

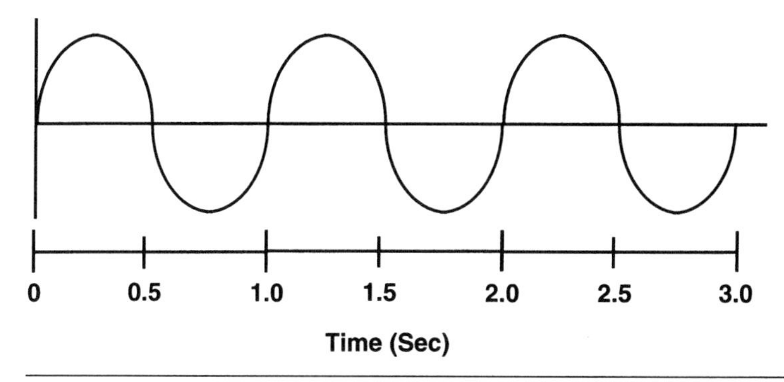

FIGURE 3-14 Period of a Sine Curve

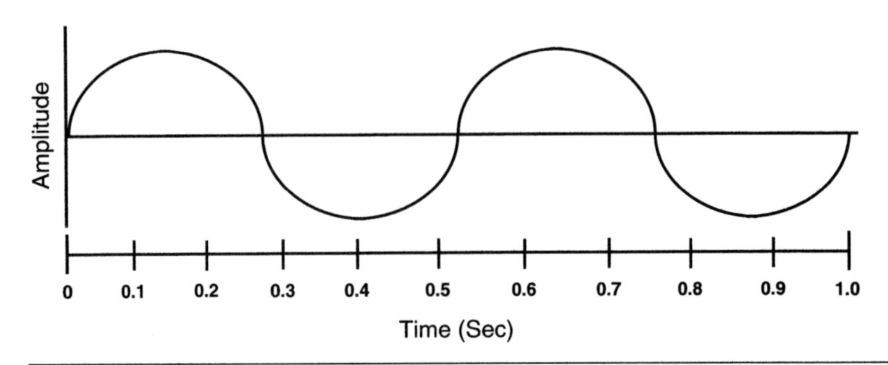

FIGURE 3-15 Frequency of a Sine Curve

Frequency

Frequency is the number of complete cycles that occur during a certain time period, usually one second (Figure 3-15). It is expressed in cycles per second (cps) or Hertz (Hz) (in honor of Heinrich Hertz, the first person to demonstrate electromagnetic waves). If the swinging pendulum or tuning fork tine or mass in the spring-mass model completes 100 cycles in one second, its frequency of vibration is 100 cps or Hz. The **pitch** of a signal is the perceptual correlate of frequency. For example, a 100-Hz pure tone would be perceived as being lower in pitch than a 1,000-Hz pure tone. As is true for loudness, pitch determination requires human perceptual judgments of the sound. In the case of human vocal fold vibrations, frequency is determined by the number of openings and closings of the vocal folds that occur in one second. If the folds open and close 100 times in a single second, the frequency of their vibration is 100 cps or Hz.

Frequency–Period Relationship

There is an inverse relationship between period and frequency. Since period is the time needed for the completion of one cycle of vibration, as frequency is increased (more cycles per second), period will be reduced (less time for the completion of any one particular cycle). Thus, as frequency is increased, period is decreased proportionately (Figure 3-16). A pure tone of 250 Hz will have a longer period (1/250 or 0.004 seconds) than one of 1,000 Hz (1/1,000 or 0.001 seconds). The reciprocal relationship between period and frequency is expressed in the following formula:

$$frequency = \frac{1}{period}$$

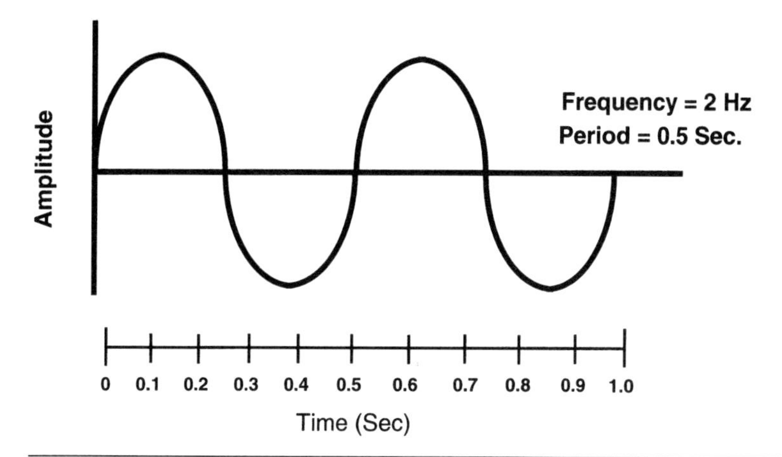

FIGURE 3-16 Reciprocal Relationship Between Frequency and Period

If the frequency of vibration for a particular sound wave is 1,000 Hz, the period would be 0.001 second (period = 1/1,000 Hz = 0.001 second).

Frequency–Wavelength Relationship

An inverse relationship exists between the time concept of frequency and the spatial concept of wavelength. As frequency is increased, wavelength becomes shorter, and as frequency is decreased, wavelength gets longer. Since the number of cycles are increased within the same unit of time (a second), each cycle will take less time and cover a shorter distance (Figure 3-17). It is an established fact in environmental acoustics that lower frequencies are more difficult to absorb than higher frequencies because of the longer wavelengths of power frequencies. A frequency of 100 Hz, for example, has a wavelength of 11 feet, while a frequency of 10,000 Hz has a wavelength of only 0.11 feet. The 10,000-Hz tone could be absorbed by acoustical ceiling tile that is only a few inches thick. However, the 100-Hz frequency would require an unusually thick wall or some other type of acoustical treatment for it to be completely absorbed.

The relationship between frequency and wavelength can be expressed in the following formulas:

$$\lambda = \frac{v}{f}$$

$$f = \frac{v}{\lambda}$$

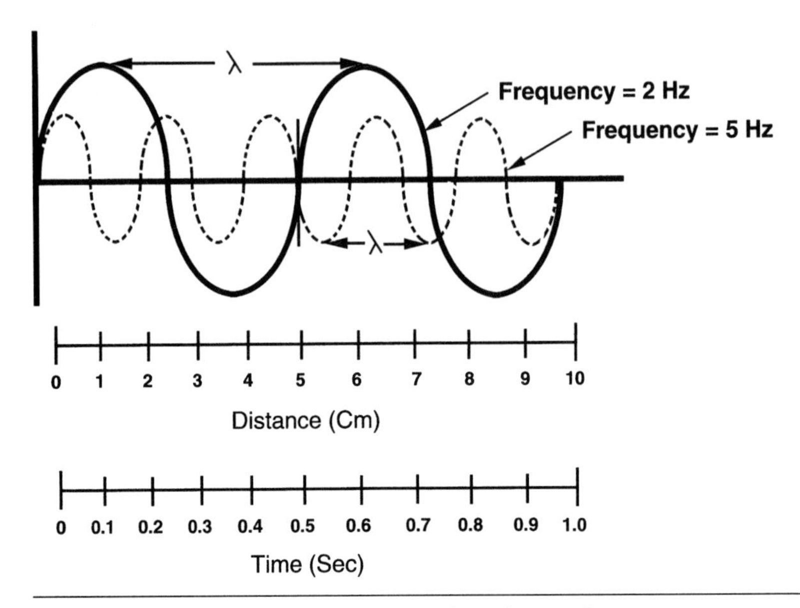

FIGURE 3-17 Frequency–Wavelength Relationship

where f is frequency, λ is wavelength, v is velocity (a constant; refers to the speed of sound). Some examples follow:

> *If the frequency of vibration for a particular sound wave disturbance is 100 Hz, the wavelength for that frequency is 11.0 feet, or 3.4 meters (wavelength = velocity/frequency: 1,100 feet per second/100 Hz = 11.0 feet; 340 meters per second/100 Hz = 3.4 meters; 34,000 centimeters per second/100 Hz = 340 centimeters).*

In this example, if the unit of measurement for velocity is feet per second, then the wavelength is expressed in feet. If the unit of measurement is meters per second, then the wavelength is expressed in meters. It is important to note that the answer is not expressed in feet or meters *per second,* but in feet or meters. Wavelength is a linear measurement of the distance covered by a sound wave disturbance during one cycle of its vibration.

> *If the wavelength for a particular sound wave disturbance is 1.1 feet, or 0.34 meters, the frequency for that sound wave disturbance would be 1,000 Hz (frequency = velocity/wavelength: 1,100 feet per second/1.1 feet = 1,000 Hz; 340 meters per second/0.34 meters = 1,000 Hz).*

Velocity

Velocity is the speed of sound through a transmitting medium. The average speed of sound in the medium of air is approximately 1,100 feet per second, or 340 meters per second, or 34,000 centimeters per second. Different sources will vary slightly with regard to these figures because there are some differences in the speed of sound in air as velocity is measured at different heights above sea level. The speed of sound in air is relatively constant because of the elastic and inertial properties of a given medium. Water has different elastic and inertial properties than does air and, consequently, the speed of sound is faster in water than in air.

In general, the velocity of sound varies as a function of the elasticity, density, and temperature of the transmitting medium, with elasticity being the most important factor. The greater the elasticity (springiness) of a medium, the greater the velocity; the greater the density of the medium (mass per unit of volume), the slower the velocity. If we think of the particles in the two media shown in Figure 3-18 as cars on a highway, then we notice that there are many fewer cars (less mass per unit of volume, lower density) in medium A than in medium B, and therefore the potential speed of movement of cars on highway (medium) A is greater than on highway (medium) B.

Temperature has an indirect effect on velocity: An increase in temperature causes a decrease in density in a medium, which, in turn, causes an increase in velocity. As proof, consider this situation: If a solid is placed in an oven, as the temperature increases, the solid will turn to a liquid and, eventually, to a gas. In progressing from a solid to a liquid and then

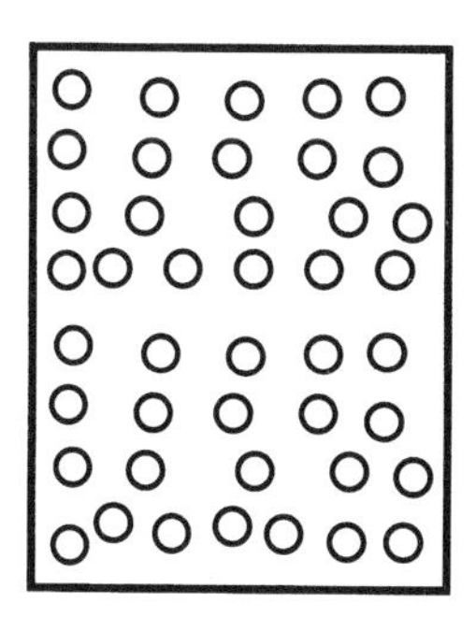

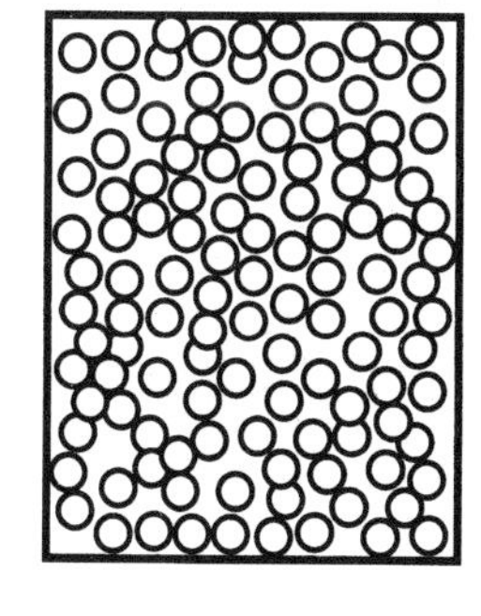

FIGURE 3-18 Media Density

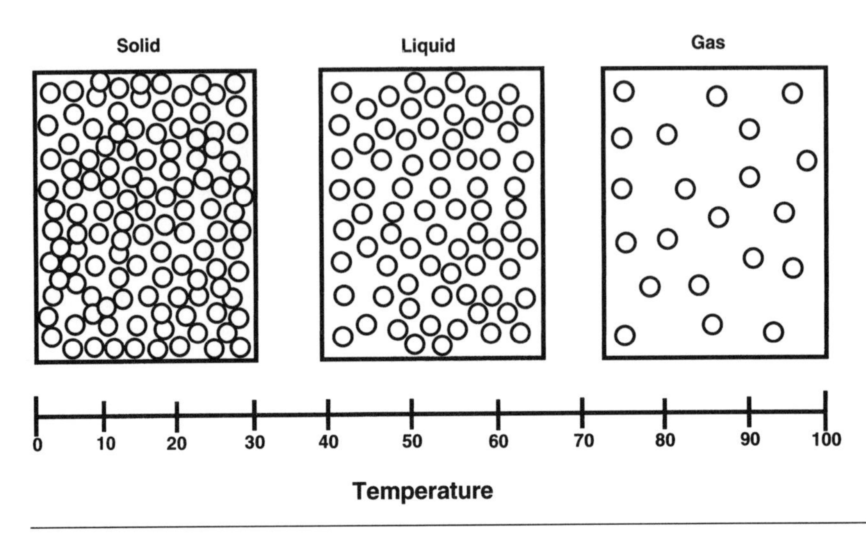

FIGURE 3-19 Temperature Effects on Density in Medium

to a gas via an increase in temperature, density has decreased as well (Figure 3-19).

Question:
Why do sound waves travel faster through the medium of steel than through the medium of air?

Answer:
Because while the density of steel (a solid) is greater than that of air (a gas), the elasticity of steel is greater than that of air, and elasticity is the primary factor in determining the velocity of sound waves.

Longitudinal versus Transverse Waves

Sound waves are **longitudinal waves;** in longitudinal waves, the particles of the medium move in the same line of propagation as the wave, that is, in the same direction as (parallel to) the movement of the wave (Figure 3-20). On the other hand, in **transverse waves,** the particles of the medium move perpendicular (at right angles) to the movement of the wave; for example, while the wave may be moving from right to left, the particles are being displaced up and down from their rest positions (see Figure 3-20). Water waves are transverse waves, a fact that becomes apparent when a rock is thrown into a pond and ripples (waves) in the

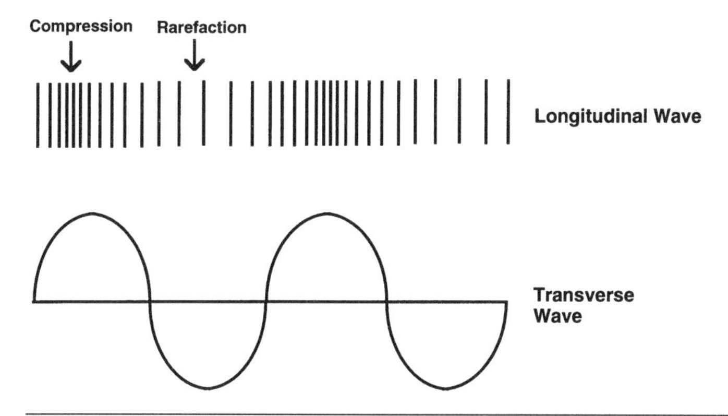

FIGURE 3-20 Longitudinal Versus Transverse Waves of the Same Wavelength

water result. While the waves are moving out in a concentric (circular) manner from the disturbance-producing rock, the water particles are moving up and down, perpendicular to the wave motion (Figure 3-21).

While transverse waves (sine curves) are used to illustrate the various properties of sound waves (e.g., amplitude, wavelength, period, cycle, etc.), in reality, sound waves are longitudinal (not transverse) waves, but their properties are more easily illustrated on transverse waves. We cannot

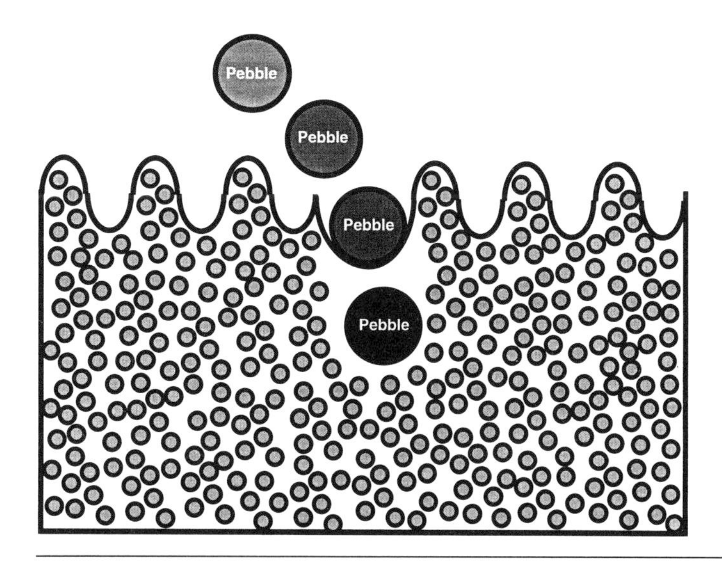

FIGURE 3-21 Perpendicular Wave Movement

see sound waves, but if we could, they would not look like the sine curves that are used to illustrate their properties.

Sound Propagation and Interference

The sound disturbance that is set up in an appropriate medium is spherically propagated through the medium. The disturbance emanates outward in all directions from the sound source until it strikes an object(s) that would alter its spherical pattern. The **inverse square law** verifies the fact that sound propagation is spherical in nature by showing the predictability of amplitude measurements at specified distances from the sound source (Figure 3-22). This law states that there is an orderly relationship between a decrease in sound amplitude and the distance that it is measured from the sound source. The amplitude of a sound at a given distance from the sound source is inversely proportional to the square of the distance of the point of measurement from the sound source. Thus, a sound of a given intensity has one-ninth (i.e., $1/(3)^2$) of its original intensity at three times the distance from a sound source.

The major assumption required for the inverse square law is that the sound wave being measured does not strike an object prior to the amplitude measurement. This law is of little practical value because the world

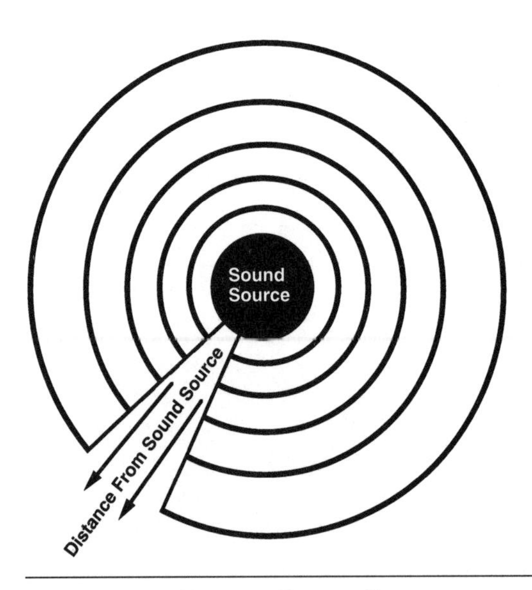

FIGURE 3-22 Inverse Square Law

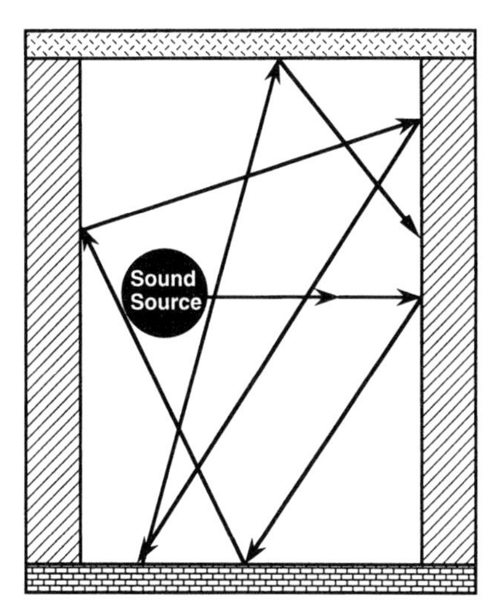

FIGURE 3-23 Reverberation

we live in has many sound "barriers," and a sound disturbance usually does not emanate too far from its originating source before it strikes an object of some kind. Once a sound wave strikes an object, several things can happen to it. The sound energy being emitted can be **absorbed** by the object that has been struck. If the object is a wall with absorptive properties, the sound energy enters the structure, is converted to thermal energy (heat), and is then dissipated. Alternatively, when sound strikes an object, it can bounce off the object. When a sound wave bounces off a wall, it is said to be **reflected.** If the reflections are multiple or continuous to the point that they actually prolong the existence of the sound within a confined space, they are referred to as **reverberations,** the prolongation of a sound through multiple or continuous reflections (Figure 3-23).

Another type of interference in the transmission of sound waves is **refraction** (or **deflection**), which is the bending of sound waves from their path of propagation as a result of changes in the determinants of velocity in the medium. Because sound waves tend to move outward from the sound source in the form of a spherical wave, after approximately six feet out from the sound source, the wavefront approaches a plane (flat) surface, and thus the wave is a **plane wave.** If, for example, temperature varies in different locations in the medium, thereby causing variations in the velocity of sound waves in that medium, then the wavefront may be tilted, thereby changing the direction of the propagation of the wave (*sound ray*) and causing a bending (refraction) of the wave from its original path

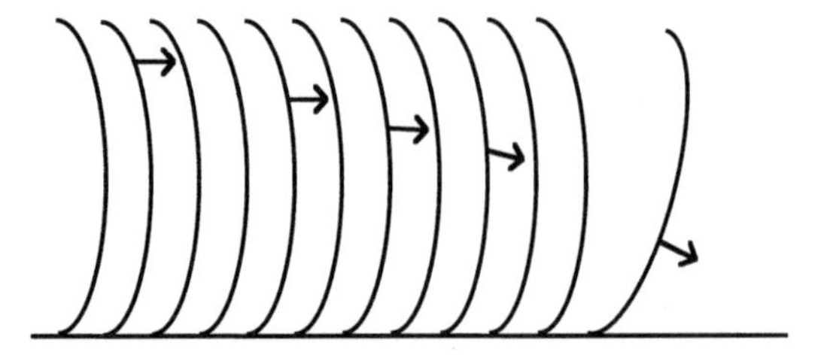

FIGURE 3-24 **Refraction of Sound Wave**

of propagation, which can cause a distortion of the sounds being transmitted (Figure 3-24). Refraction effects can be seen in rooms with very tall ceilings in which the temperature varies at different room locations (e.g., warmer on the ceiling than on the floor), leading to differences in the intelligibility of sounds at different locations in these rooms.

Complex Sounds

So far our discussion of basic acoustics has centered around simple sound disturbances. These sinusoidal disturbances have been shown graphically on an amplitude-by-time display known as a **waveform** graph (Figure 3-25). For simple harmonic motion, the waveform graph clearly displays amplitude changes as a function of time. Another method for displaying sound is to graph it in terms of amplitude as a function of frequency.

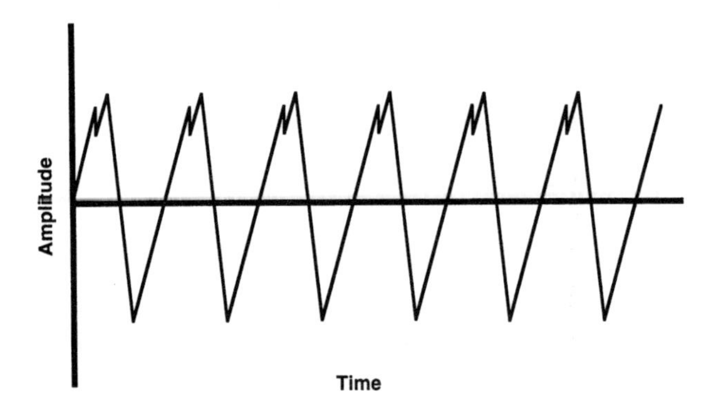

FIGURE 3-25 **Complex Waveform**

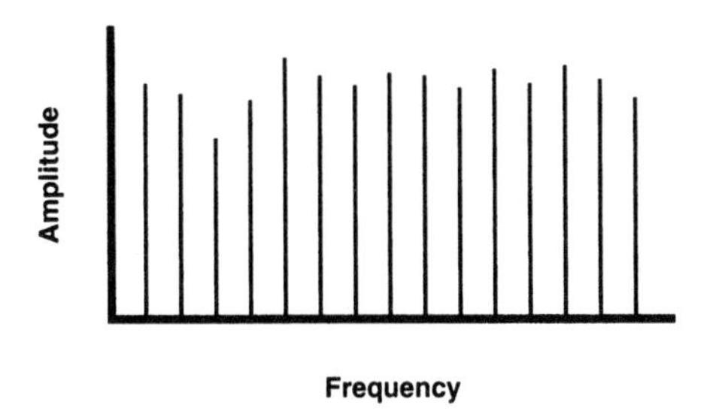

FIGURE 3-26 Spectrum of a Complex Periodic Sound

When amplitude is plotted as a function of frequency, the resulting graph is referred to as a **spectrum** (Figure 3-26). A spectrum for simple sounds (pure tones) would consist of a single line located at the appropriate frequency (Figure 3-27). The vertical length of the single line would be equal to the amplitude of the pure tone that has been graphed. A spectrum shows amplitude as a function of frequency at a single instant in time, and has the advantage of allowing frequency to be read directly from the display. A waveform has the advantage of showing amplitude changes over time, but frequency has to be calculated. The spectrum provides little advantage when viewing simple sound disturbances, because all of the energy is concentrated at a single frequency. However, when viewing

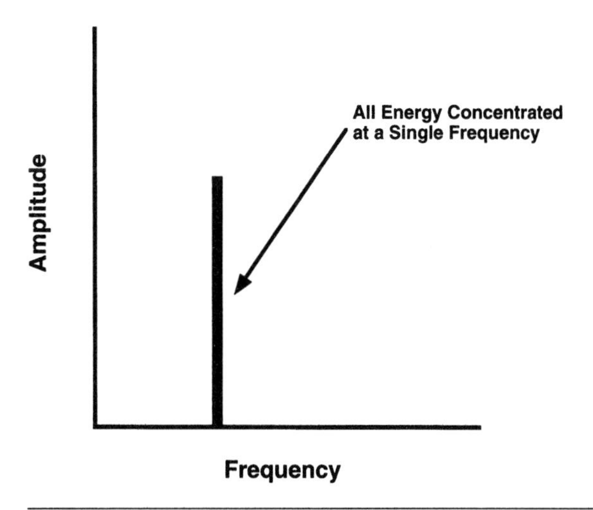

FIGURE 3-27 Spectrum of a Pure Tone

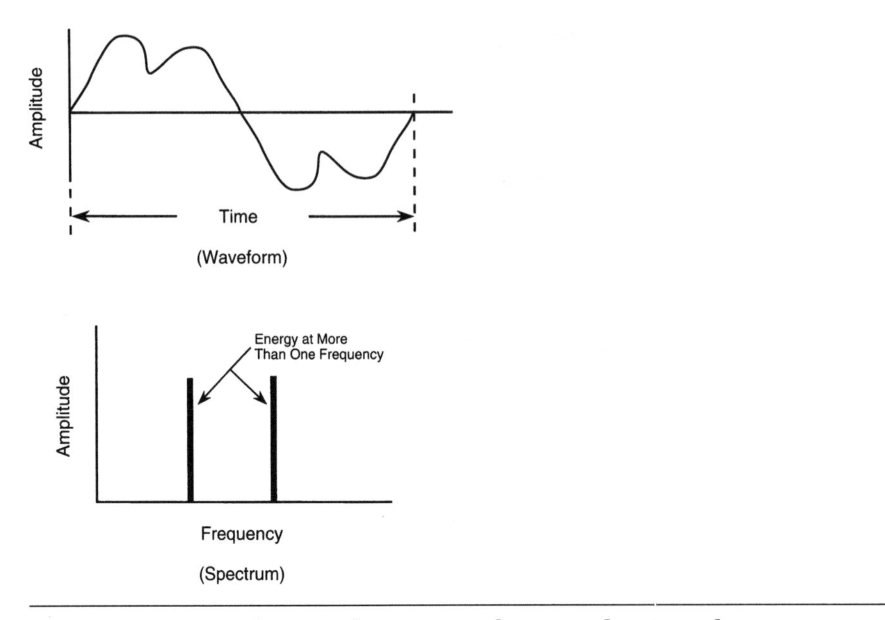

FIGURE 3-28 Waveform and Spectrum for Complex Sound

complex sounds in which there is energy at more than one frequency, the sound spectrum becomes more valuable.

Complex sounds differ from simple sounds in that they have energy distributed at more than one frequency. A single tuning fork generates a sound with energy concentrated at one frequency. If two tuning forks of different frequencies are activated simultaneously, the sound generated will consist of two frequencies and will therefore be considered complex in nature. The resultant waveform will no longer show smooth

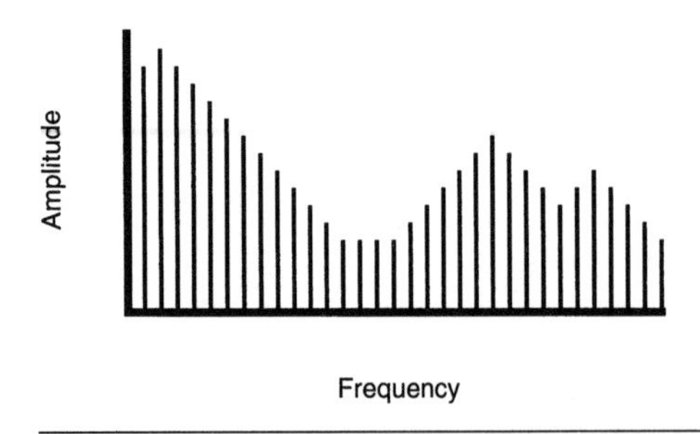

FIGURE 3-29 Spectrum for /i/

curves like that of the sine wave, and the spectrum will have two vertical lines, each line representing the frequency of vibration of one of the tuning forks vibrating simultaneously with the other (Figure 3-28). Speech sounds, like those for the vowels of English, are very complex in that they have energy distributed at numerous frequencies, with amplitude variations at each of the frequencies involved. Figure 3-29 shows the sound spectrum for the vowel /i/ as in the word b**ee**t.

Periodicity versus Aperiodicity

A periodic sound disturbance is one in which the wave shape repeats itself as a function of time; that is, the wave shape is said to have **periodicity** (Figure 3-30). A pure tone that provides simple harmonic motion is, by definition, periodic. So is the swing of a pendulum or tuning fork tine. The pure tone has a clearly defined frequency because of the cyclical (periodic) behavior of the vibrator generating it. While the vowels of English are not simple sounds like the pure tone, they are periodic because of the cyclical nature of the sound generator (the human vocal folds) employed during their production. The human vocal folds open and close in a rhythmic manner during the production of vowels, causing repetitive (technically quasiperiodic) wave shapes to occur (Figure 3-31).

J.B. Fourier, a French mathematician who lived in the early part of the nineteenth century, showed that any complex periodic sound wave disturbance can be mathematically broken down into its individual sine wave (pure tone) components, which vary in terms of frequency, ampli-

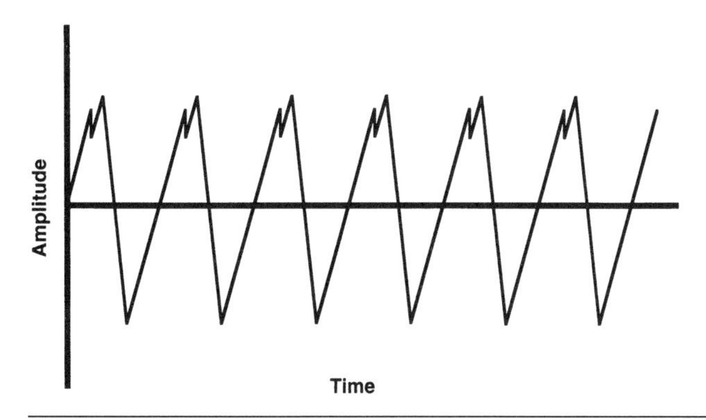

FIGURE 3-30 Periodic Waveform

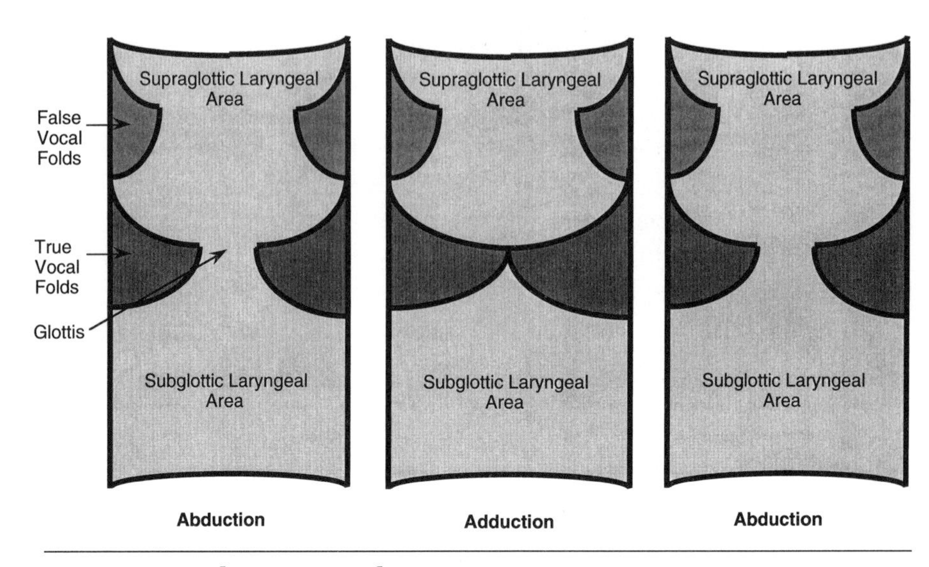

FIGURE 3-31 Phonatory Cycle

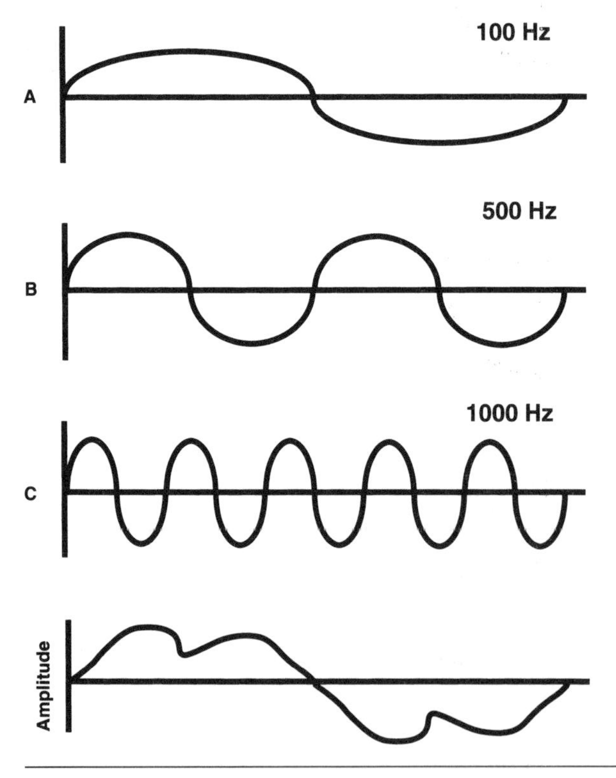

FIGURE 3-32 Fourier Analysis

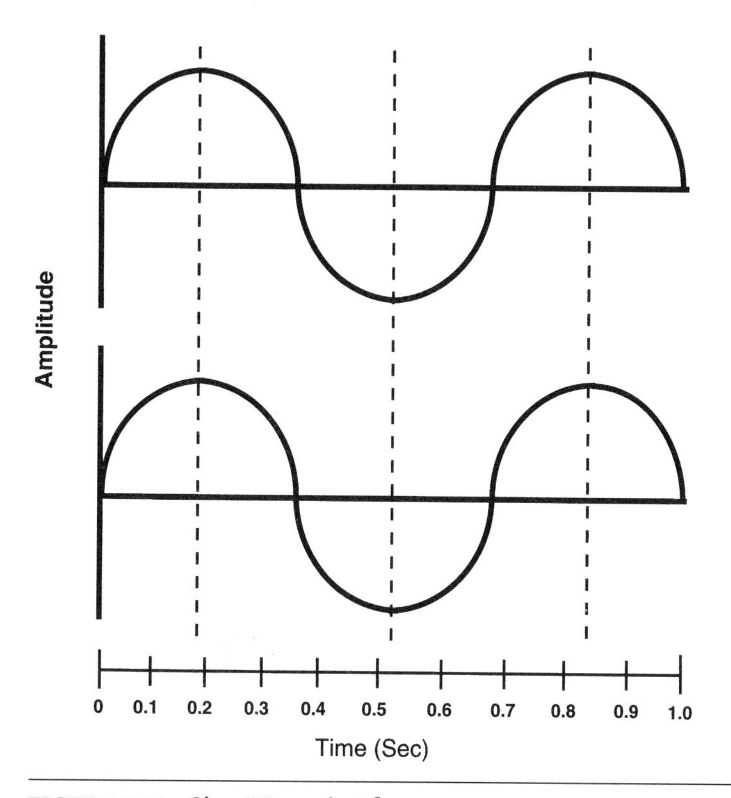

FIGURE 3-33 Sine Waves in Phase

tude, and/or phase relations with respect to one another (Figure 3-32). This mathematical analysis of complex signals into their sinusoidal components is called **Fourier analysis.** (Frequency and amplitude have already been defined.) **Phase** is the portion of a cycle through which a vibrator has passed up to a given instant in time; it is concerned with the timing relationships between individual sinusoids. Two sinusoids are *in phase* when their wave disturbances crest and trough at the same time (Figure 3-33) and *out of phase* when they do not (Figure 3-34). The human ear does not seem to detect phase differences or to use them to any great degree in the interpretation of the speech signal. If frequency, amplitude, and phase are all considered together, Fourier analysis can be used to determine the sine waves that are combined to produce any complex periodic sound disturbance.

An aperiodic sound disturbance is one in which the wave shape does not repeat itself as a function of time, and is therefore said to have

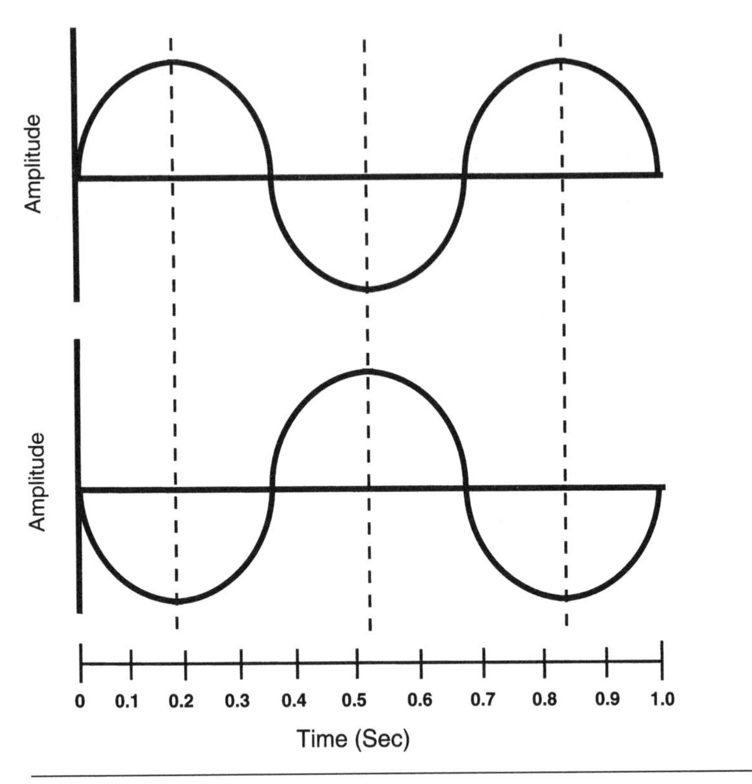

FIGURE 3-34 Sine Waves Out of Phase

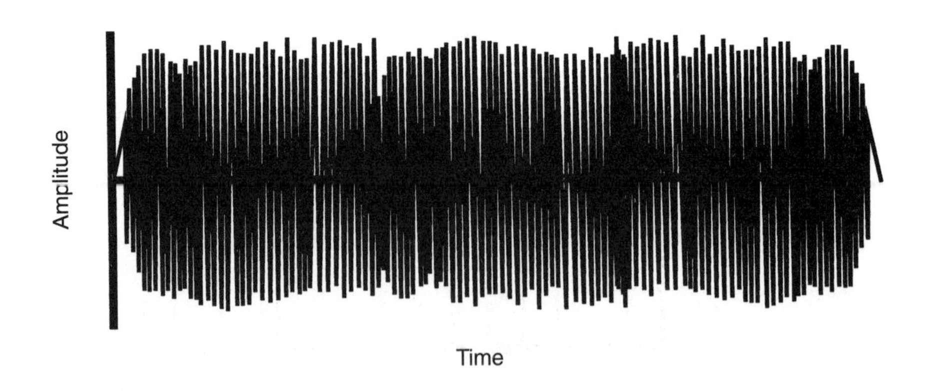

FIGURE 3-35 Aperiodic Waveform

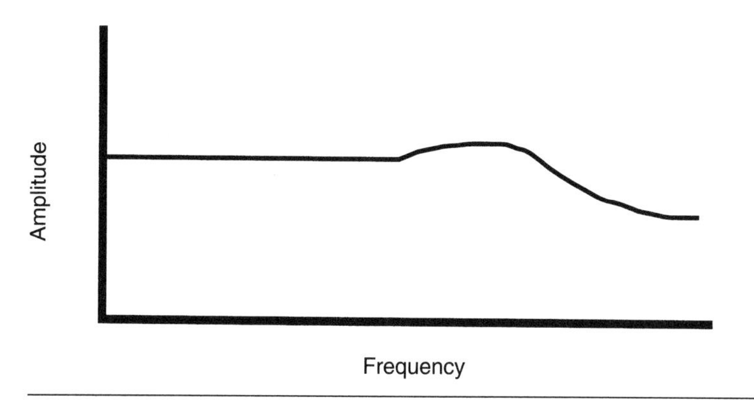

FIGURE 3-36 Spectral Envelope for a Complex Aperiodic Sound (/s/)

aperiodicity (Figure 3-35). Static on the radio and a sudden explosion are examples of an aperiodic sound disturbance. When these sounds are heard, they are usually perceived as noise, because they lack any cyclical or repetitive vibrations. A voiceless fricative sound such as /s/ (as in **s**un) would be an example of a complex aperiodic speech sound. The phoneme /s/ is produced by forcing air through a narrow constriction within the oral cavity. Consequently, there is no cyclical or repetitious activity to be heard, and the result is noise-like to the ears. The wave shape for /s/ does not repeat itself over time (i.e., it is aperiodic) (Figure 3-36).

Spectral displays of complex periodic and complex aperiodic sounds reveal the major differences between them. For complex periodic waves, the frequency of each component is a whole-number multiple of the component with the lowest frequency, called the **fundamental.** Figure 3-37 shows a spectrum of a complex periodic sound. The first bar (i.e.,

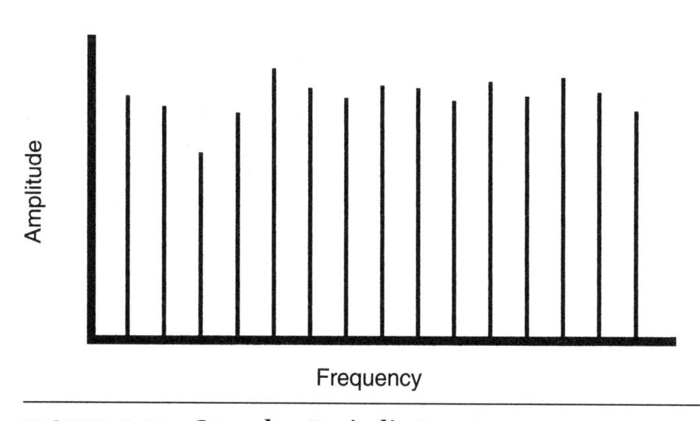

FIGURE 3-37 Complex Periodic Spectrum

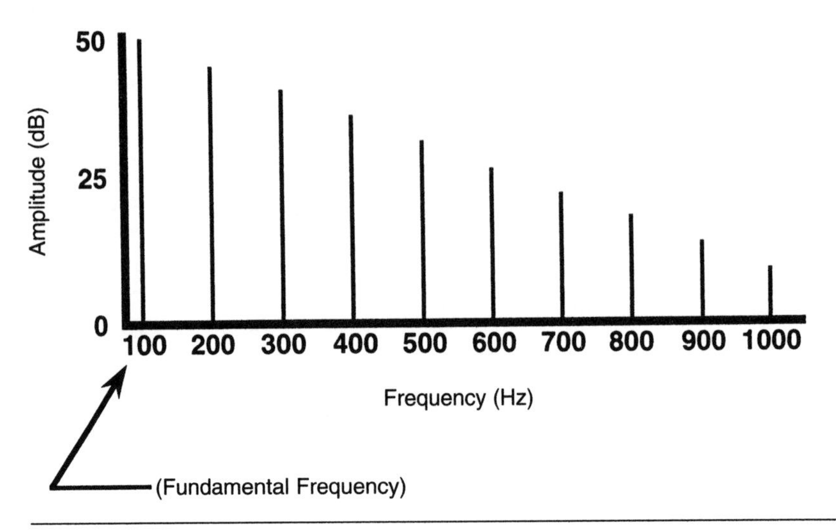

FIGURE 3-38 Spectrum Showing Fundamental Frequency and Harmonics

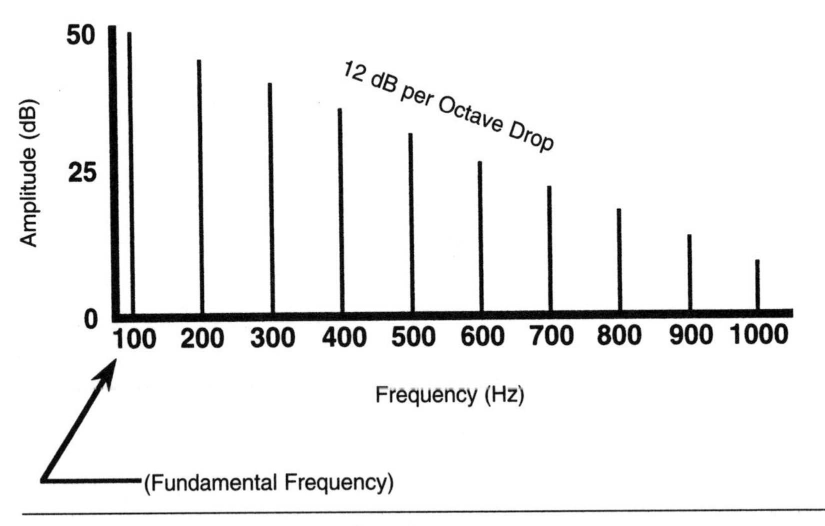

FIGURE 3-39 Spectrum of Glottal Source

the bar showing the lowest frequency) is the fundamental frequency, and the energy bars above it are whole-number multiples of the fundamental frequency. If the lowest bar of energy has a frequency of 100 Hz (Figure 3-38), then the second energy bar has a frequency of 200 Hz, the third has one of 300 Hz, and so on. The spectrum shown in Figure 3-39 is that of the sound wave disturbance that would be derived from vibration of the human vocal folds. In this spectrum, the lowest (fundamental) frequency is 100 Hz, and the other energy bars, called **harmonics,** have frequencies that are whole-number multiples of the fundamental frequency. The heights of the energy bars for the various frequencies in this spectrum refer to the relative amplitude for each sinusoid making up this complex periodic sound. In the case of a sound produced by the human vocal folds, there is a uniform energy drop with an increase in frequency. In this spectrum, the pure tone component with the highest concentration of energy (i.e., the greatest amplitude) is the fundamental, the component pure tone in this complex signal with the lowest frequency. For complex aperiodic sounds, there is no fundamental frequency or harmonics, because the disturbances produced do not set up any cyclical or repetitious behavior. Instead, there is energy distributed throughout the sound spectrum at a particular instant in time.

The top illustration in Figure 3-40 shows the spectrum for **white noise,** which sounds like the hissing of a radiator. White noise has energy distributed evenly throughout the spectrum and is therefore useful for masking other sounds. Instead of having discrete lines (representing concentrations of energy or energy bars), like those used for complex periodic sound spectra, a device called a **spectral envelope** is employed to show the distribution of energy for complex aperiodic sound disturbances. The spectral envelope is a line running horizontally across the spectrum, which, in this case, because it is a flat line, indicates that there is energy distributed evenly throughout the frequency range. If the spectrum is showing an aperiodic signal other than white noise, the spectral envelope would not be completely flat, but would show variations where higher or lower energy regions within the frequency range would be located.

The lower illustration in Figure 3-40 shows the spectrum for the speech sound /s/ (as in **s**un). In this instance, there is a concentration of energy in the higher frequency range, which is shown by a "rise" in the spectral envelope for the frequencies where the energy concentration is located.

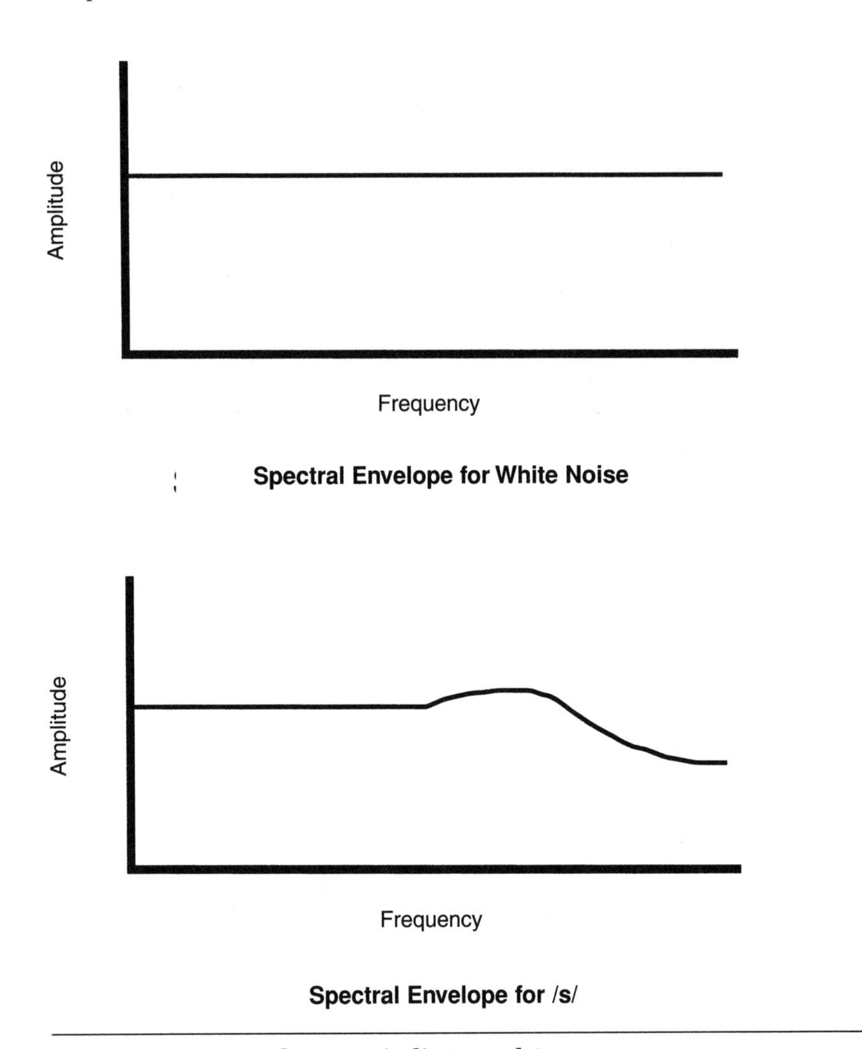

FIGURE 3-40 Complex Aperiodic Sound Spectra

Sound Visualization

Up to this point, two methods for graphically representing sound have been discussed. The waveform shows sound wave disturbances in terms of amplitude variations as a function of time, with amplitude on the vertical axis and time on the horizontal axis (Figure 3-41, top illustration). The spectrum shows amplitude variations as a function of frequency, with amplitude displayed on the vertical axis (just as it is on a waveform),

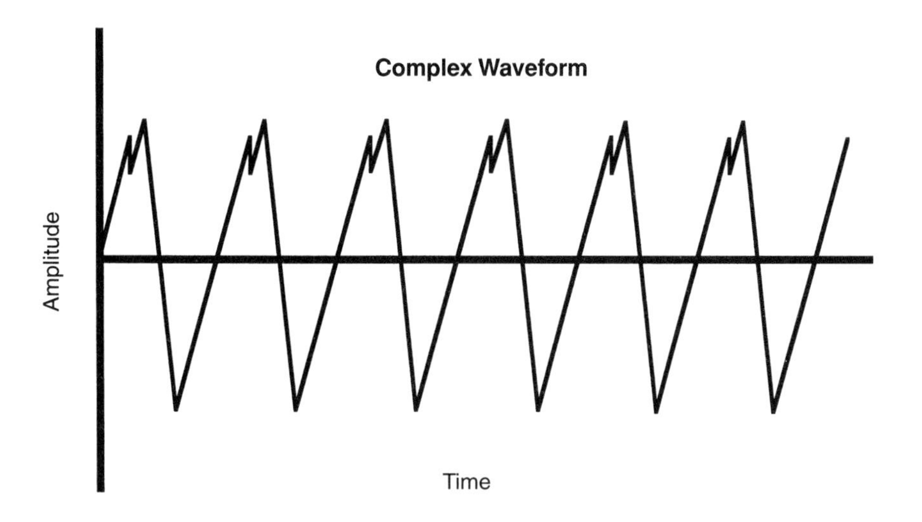

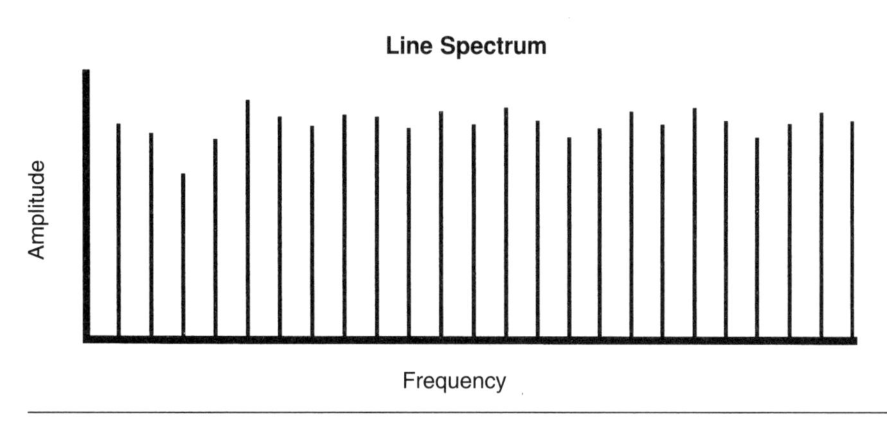

FIGURE 3-41 Complex Waveform and Line Spectrum

but the horizontal axis shows frequency instead of time (Figure 3-41, bottom illustration). The spectrum, therefore, shows sound at a single instant in time, and the waveform shows sound amplitude variations over time.

The speech signal is a constantly changing, ongoing signal that contains many frequency and amplitude variations occurring over time. To study the frequency and amplitude changes that occur during speech production, speech scientists have developed the *sound spectrograph,* a

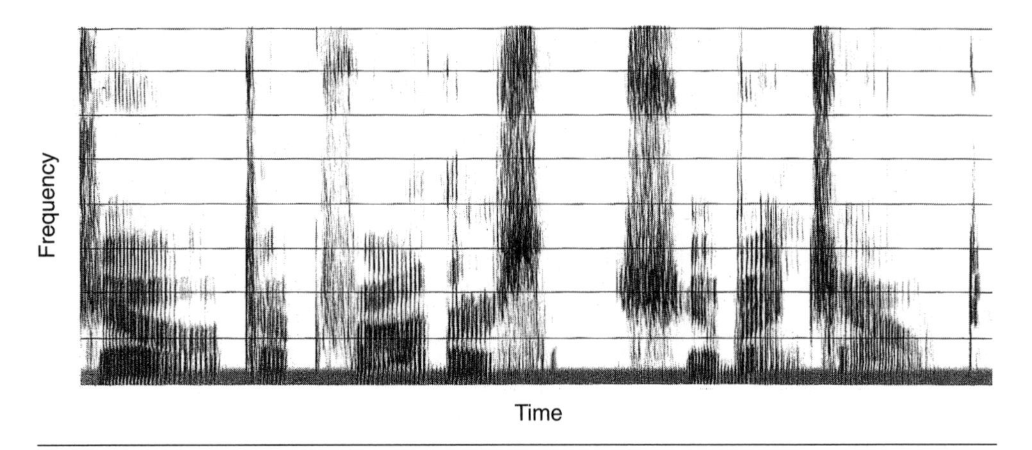

FIGURE 3-42 Spectrogram

device that can analyze, by means of a series of analyzing filters, the speech signal and produce **spectrograms,** which show disturbances in terms of frequency and amplitude variations as a function of time (Figure 3-42). Spectrograms display frequency on the vertical axis and time on the horizontal axis, while amplitude variations are shown as varying degrees of darkness within the spectrogram itself (the darker areas in Figure 3-42 are regions of higher concentrations of energy). While spectra show amplitude variations as a function of frequency, spectrograms provide the same information as well as information on the temporal dimension of speech.

The Decibel

While the amount of energy in individual speech sounds is very small, the range of audible energy, from the sound with the least energy to the sound with the most, is very large. The decibel scale avoids the cumbersomeness of dealing with numbers in linear, absolute scales, like power and pressure. In linear scales, all units of measurement are the same size; for example, in a linear scale of distance, the distance between 2 and 3 inches on a ruler is exactly the same as between 50 and 51 inches. This is not true of logarithmic scales, in which each unit is larger than the preceding unit. In addition, unlike absolute scales of measurement with a true zero point, the decibel scale is a relative scale in which there is an arbitrary zero point used as a reference point. Therefore, the decibel scale expresses the ratios between two sound pressures or powers or the equivalent of a given power or pressure measurement relative to a reference point. It is an effective means of expressing sound amplitudes because it is much more convenient than using absolute units of measurement; it provides a much smaller, more manageable range of numbers. For example, the range of energy for most audible sounds runs from –10 to 140 dB (threshold of pain). The key to making this conversion to the smaller range of numbers lies in the use of logarithms.

Logarithmic scales use logarithms or exponents, which are values to which a base must be raised in order to equal a desired number. In the decibel scale, the base is 10. Table 3-1 shows the logarithm to the base 10 for a series of numbers. The logarithm (to the base 10) of the number 100,000 is 5; this means that 10 X 10 = 100 X 10 = 1,000 X 10 = 10,000 X 10 = 100,000.

Thus, the numbers 10, 100, 1,000, 10,000, and 100,000 can be represented by the following set of numbers: 10^{-3}, 10^{-2}, 10^{-1}, 10^{0}, 10^{1}, 10^{2}, 10^{3}, 10^{4}, 10^{5}, 10^{6}. The superscript number (called an **exponent** or **logarithm**) in the scale tells us how many times 10 must be multiplied by itself to obtain the desired number. For example, 10^2 indicates that 10 X 1 = 10, X 10 = 100, which is the equivalent of 10^2; 10 to the third power (10^3) shows that 10 X 1 = 10, X 10 = 100, X 10 = 1,000, which is the ratio scale equivalent of 10^3. The number 10, then, is the base that must be raised to some power in order to represent larger numbers on a ratio scale. The exponents to which the base is raised are logarithms. The range of numbers from 10 to 1,000,000 can be represented by the smaller set of logarithms ranging from 1 to 5, with the assumption being that the base for the interval scale is 10.

TABLE 3-1 *Logarithms (Exponents) to the Base of 10*

Desired Number	*Base 10*	*Logarithm*
10	10	1
100	10 X 10	2
1,000	10 X 10 X 10	3
10,000	10 X 10 X 10 X 10	4
100,000	10 X 10 X 10 X 10 X 10	5
1,000,000	10 X 10 X 10 X 10 X 10 X 10	6

If this scale represents sound amplitudes, then the logarithms (exponents) in the scale can be assumed to be the number of *Bels* representing that ratio of sound amplitudes. The Bel is an arbitrary logarithmic unit used to represent sound amplitude on an interval scale. The term *Bel* was derived in honor of Alexander Graham Bell (1847–1922). (Although Alexander Graham Bell is most famous for the invention of the telephone, he was also a deaf educator and created one of the earliest symbol systems for teaching the deaf to communicate.)

The decibel is one tenth of a Bel. If one Bel is equal to 10 decibels, then the range of human hearing can be represented by 140 dB, which seems to be a convenient interval scale size for expressing sound amplitudes.

Since the decibel is a relative unit used to express sound amplitudes, it can be conveniently applied to a number of scales that are routinely used for making amplitude measurements. One such scale is the *power scale,* which measures amplitudes in terms of watts/cm^2. The arbitrarily chosen reference point for this scale is 10^{-16} watts/cm^2; this reference point is equivalent to one of the weakest sound amplitudes that human ears can detect. The formula for expressing decibel ratios when using the intensity level scale is

$$dB = 10 \log R \, (I_1/I_2)$$

where 10 is a constant for transforming the Bel into decibels (there are 10 dB in one Bel) and R is the ratio of the intensity of one sound (I_1, the sound with the greater intensity) to the intensity of the other sound (I_2, the sound with the lesser intensity). Some examples of how this formula works are as follows:

1. *If the amplitude of a sound is 1,000 times greater than another sound's amplitude, the decibel difference between these two sounds is 30 dB (dB = 10 X log 1,000/1; 1,000 divided by 1 = 1,000. The logarithm of 1,000 (to the base 10) = 3; 10 X 3 = 30 dB).*
2. *If one sound amplitude is 50 watts/cm^2, and another sound amplitude is 5 watts/cm^2, the decibel difference for these two sounds is 10 db*

$$dB = 10 \times \log \frac{50 \text{ watts/cm}^2}{5 \text{ watts/cm}^2} = 10;$$

the logarithm of 10 (to the base 10) = 1; 10 X 1 = 10 dB.

3. *If the amplitude of a sound is 150,000 watts/cm², its intensity in decibels is*

$$10 \times \log \frac{150{,}000\ watts/cm^2}{10^{-16}\ watts/cm^2}$$

The logarithm of a number can be determined through the use of log tables, which can be found in most statistics books, or through the use of a scientific calculator, which usually has a "log" key on it. It is assumed that the log key on the calculator represents a log scale to the base 10, unless specified otherwise. In the preceding examples, the logarithm of the numbers involved could be determined by counting zeros. The log (to the base 10) of 1,000 is 3; the log (to the base 10) of 10 is 1, and so on.

Sound amplitudes are usually determined by the amounts of pressure (force per unit area) that they are causing to occur on a thin plate or diaphragm. The scale most often used to measure sound amplitude is the **sound pressure level** scale. This scale measures amplitudes in terms of $dynes/cm^2$ or Pascal (Pa). The reference point for this scale is 0.0002 $dynes/cm^2$ (20 μPa), which is roughly equivalent to the weakest sound amplitude that humans can detect. The formula for expressing decibel ratios when using the sound pressure level scale involves multiplying the logarithm of the sound amplitude ratios by 20 (instead of 10, as was used for expressing decibel ratios in power levels). The reason for this change is because intensity is equal to pressure squared ($I = P^2$). The formula for expressing decibel ratios when using the sound pressure level scale is

$$dB = 20 \log R \left(\frac{P_1}{P_2}\right)$$

where 20 is a constant; R is the ratio of the higher sound pressure level divided by the lower sound pressure level; P_1 is higher sound pressure; and P_2 is lower sound pressure. Some examples of how this formula works follow:

1. *If one sound amplitude is 1,000 times greater than another sound amplitude, the decibel difference on the sound pressure level scale is 60 dB SPL (dB SPL = 20 × log 1,000/1; 1,000 divided by 1 = 1,000. The logarithm of 1,000 (to the base 10) = 3; 20 × 3 = 60 dB SPL).*
2. *If one sound amplitude is 50 dynes/cm² (Pa) and another sound amplitude is 5 dynes/cm² (Pa), the dB difference on the sound pressure level scale is 20 dB SPL.*

$$dB\ SPL = \frac{50\ dynes/cm^2\ (Pa)}{5\ dynes/cm^2\ (Pa)}$$

50 divided by 5 = 10; the logarithm of 10 (to the base 10) = 1; 20 × 1 = 20 dB SPL.

3. *If the amplitude of a sound is 150,000 dynes/cm² (Pa), its intensity in decibels is*

$$20 \times log\ \frac{150{,}000\ dynes/cm^2\ (Pa)}{0.0002\ dynes/cm^2\ (Pa)}$$

When testing hearing, audiologists use a measurement scale that is an offshoot of the sound pressure level scale. They use the **hearing level** or **hearing threshold level** scale. The zero point for this scale is elevated from the bottom of the sound pressure level scale (0.0002 dynes/cm^2 [Pa]). It represents normal human hearing and can be different for each frequency tested because human hearing does not necessarily show the same sensitivity for each frequency. For example, at 1,000 Hz, 0 dB HL is set at approximately 7.5 dB SPL. At this frequency, the average normal listener would just begin to hear a pure tone at 7.5 dB SPL. Audiologists have developed the hearing level scale because a number of individuals have better than "normal hearing." A child, for example, who has had little exposure to loud sounds or extreme environmental noises, could have a hearing level value of –5 dB HL at 1,000 Hz.

Summary

This chapter has discussed basic acoustics, including the conditions necessary to create sound, properties of vibrating systems, simple harmonic motion, sine curves and their spatial and temporal features, longitudinal versus transverse waves, Fourier analysis, the spectral analysis of complex periodic and aperiodic sounds, and the decibel. An understanding of the physics of sound allows the reader to apply basic acoustics concepts specifically to the acoustics of speech production, which is addressed in detail in Chapter 5.

Study Questions

1. Define *sound.*
2. What are the properties of a *medium* that are necessary for the transmission of sound waves?
3. What is a *sine curve,* and why is it used to illustrate the properties of sound waves?
4. Use sine curves to illustrate *particle pressure, particle velocity,* and *particle displacement.*
5. Use sine curves to illustrate the following properties of sound waves:
   a. amplitude
   b. cycle
   c. frequency
   d. period
   e. wavelength
   f. phase
6. Using both formulas and drawings of sine curves, illustrate the relationship of
   a. frequency and period
   b. frequency and wavelength
7. What is the *frequency* of a wave traveling through the medium of air with a wavelength of 5.5 meters?
8. What is the *wavelength* of a 500-Hz sound wave traveling through air?
9. What is the *period* of a 1,500-Hz sound wave in air?
10. The following illustration shows a sound wave traveling through air.

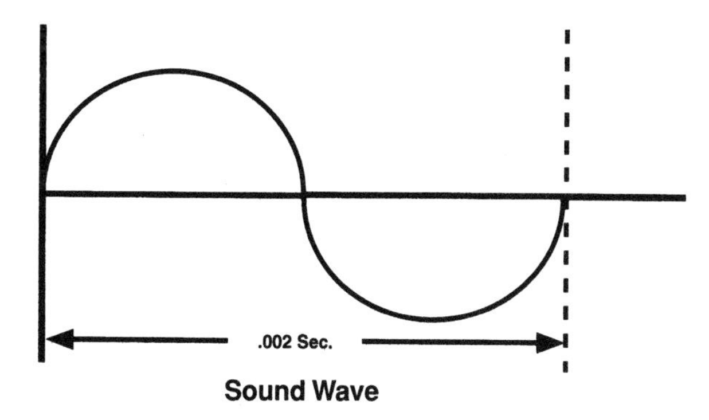

Sound Wave

Determine the following properties of this wave:
a. frequency
b. period
c. wavelength

11. Why do sound waves travel faster through the medium of steel than through the medium of air?
12. Define and differentiate *longitudinal* and *transverse waves.*
13. Define and differentiate *reflection, absorption,* and *refraction.*
14. Define and explain the *inverse square law.*
15. If sound "A" is 500 dynes/cm^2 (Pa) and sound "B" is 50,000 dynes/cm^2 (Pa), what is their intensity difference in *decibels*?
16. Sound "Z" is 20,000 watt/cm^2. What is its intensity in *decibels*?

Suggested Readings

Borden, G.J., Harris, K.S., & Raphael, L.J. (1994). *Speech Science Primer: Physiology, Acoustics, and Perception of Speech* (Third Edition). Baltimore: Williams & Wilkins.

Durant, J.D., & Lovrinic, J.H. (1995). *Bases of Hearing Science.* (Third Edition). Baltimore: Williams & Wilkins.

Kent, R.D. (1997). *The Speech Sciences.* San Diego: Singular Publishing Group.

Speaks, C.E. (1996). *Introduction to Sound: Acoustics for the Hearing and Speech Sciences* (Second Edition). San Diego: Singular Publishing Group.

Yost, W.A. (1994). *Fundamentals of Hearing: An Introduction* (Third Edition). San Diego: Academic Press.

4

Resonance

It is important to understand the concept of resonance because it relates to the articulatory aspects of speech production and will thus aid in understanding the acoustics of speech production (which is discussed in the next chapter).

If a vibrating periodic source of energy is used to activate an elastic system, the elastic system will respond by trying to vibrate at the frequency or frequencies generated by the periodic energy source. If the frequency or frequencies of the periodic energy source are similar or identical to those of the elastic system being activated, an increase in overall amplitude of vibration will occur. This phenomenon is known as **resonance.** In other words, resonance is the phenomenon whereby a body, which has a natural tendency to vibrate at a certain frequency(ies) (its natural or resonant frequencies), can be set into vibration by another body whose frequency(ies) of vibration is/are identical or very similar to the resonant or natural frequency(ies) of vibration of the first body. The closer the natural frequency(ies) of the first body to the natural frequency(ies) of the second body, the greater the amplitude of the first body's vibrations.

Thus, resonance occurs when an elastic system is forced by an external periodic source of energy to vibrate at its own natural frequency(ies) of vibration. The principle of resonance can be applied to: (1) sympathetic vibration, (2) the sounding-board effect, and (3) cavity (acoustical) resonance.

Sympathetic Vibration

All systems that can vibrate, whether simple or complex, have certain frequencies at which they vibrate best. One of the simplest of the vibrating

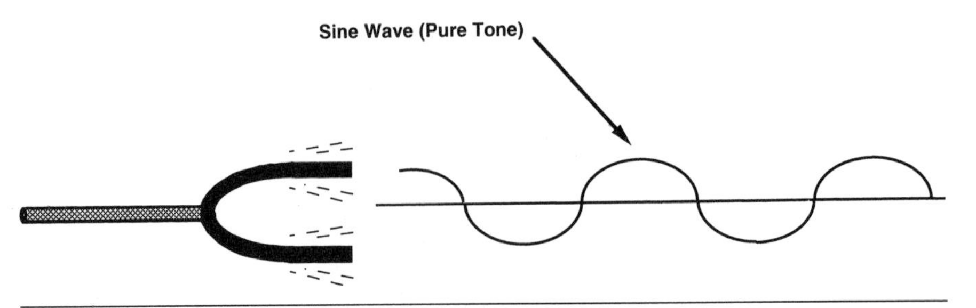

FIGURE 4-1 Tuning-Fork Generation of Pure Tones

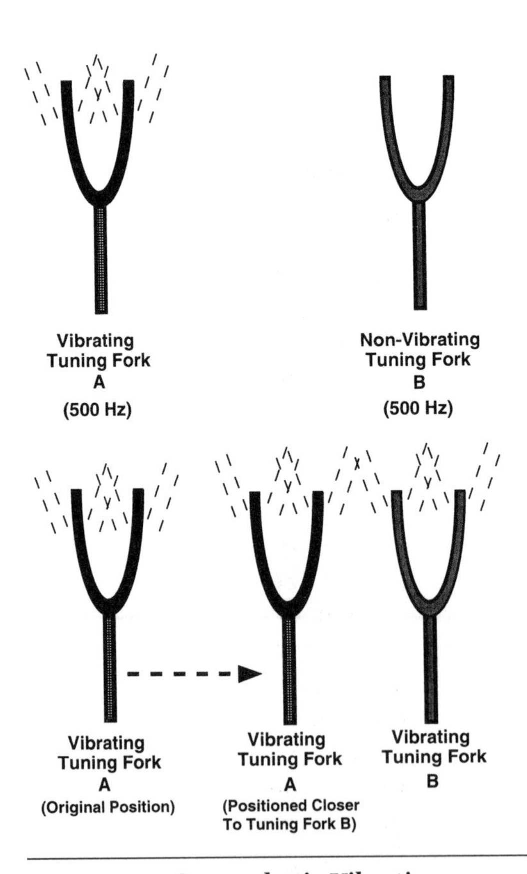

FIGURE 4-2 Sympathetic Vibration

systems is the tuning fork. The tuning fork is a mechanical device used to generate pure tones (Figure 4-1). It has two tynes (prongs), a handle, and a natural frequency of vibration (which generates a pure tone). When struck, the tuning fork will begin to vibrate (the tynes move back and forth) at its natural frequency of vibration. For example, a 500-Hz tuning fork will have a natural frequency of vibration (that frequency at which it vibrates best) of 500 Hz. If a 500-Hz tuning fork that has been struck (and thereby set in motion) is placed near a second unstruck 500-Hz tuning fork, which has not been set in motion, the second tuning fork will begin to vibrate in sympathy with the first tuning fork because its natural frequency matches that of the first fork (Figure 4-2). The result will be an increase in overall amplitude because of the occurrence of resonance. The second, unstruck tuning fork is set into **sympathetic vibration** by the vibrations of the first (struck) tuning fork.

However, if a vibrating 500-Hz tuning fork is placed near a second unstruck tuning fork with a natural frequency of 1,000 Hz, the two tuning forks are not similar in frequency and the result will be no vibration of the second tuning fork because its natural frequency of vibration is *not* the same as that of the first tuning fork.

Sounding-Board Effect

Another application of the principle of resonance is the sounding-board effect. If a struck, vibrating tuning fork is placed on a large resilient surface, it can cause a second unstruck tuning fork with the same natural frequency placed on the same surface to be set into vibration (Figure 4-3). An example of a large resilient surface is the sounding board of a piano (hence the origin of the term **sounding-board effect**); the strong tones of a piano are made possible by the action of the strings on the piano's sounding board.

Cavity (Acoustical) Resonance

Cavity (or acoustical) resonance is of particular importance to a discussion of the acoustic aspects of speech production because the human vocal tract (which extends from the glottis to the lips) can be viewed as a cavity or tube that is closed at one end (the larynx) and open at the other end (the lips) (Figure 4-4). Tubes have specific resonance characteristics that are critical to the articulatory aspects of the speech production process.

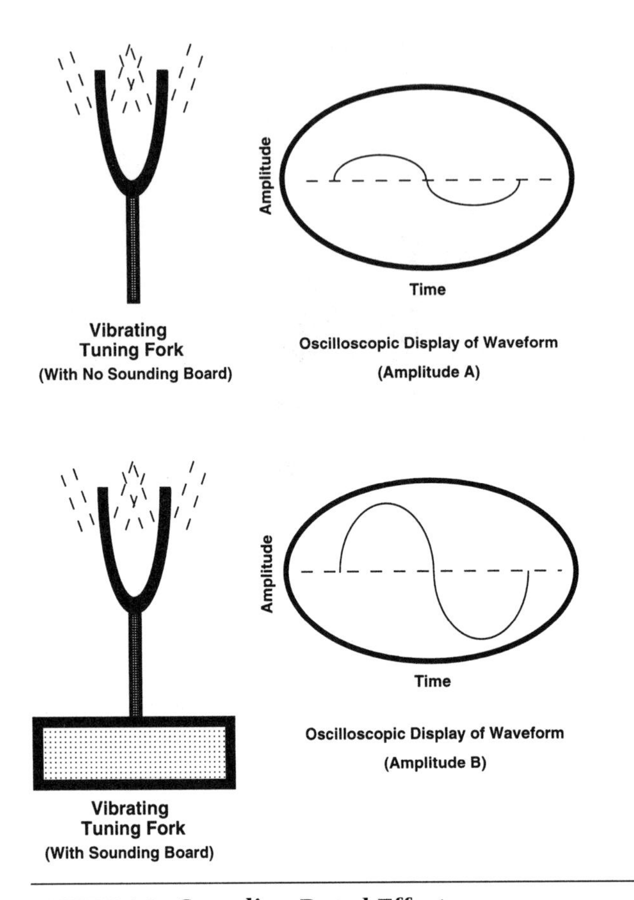

FIGURE 4-3 Sounding-Board Effect

A standard laboratory demonstration of cavity resonance consists of inserting one end of a straight tube that is open at both ends into a beaker of water (Figure 4-5). The end of the tube inserted into the water can be viewed as closed, while the unsubmerged end is considered open. The next step in the demonstration is to place a vibrating tuning fork over the open end of the tube. As the tube is slowly moved up and down in the water, a certain length of tube above the water (its *effective length*) will be found that causes an increase in the perceived amplitude of the tuning fork's tone. At this point, the length of tube above the water provides the tube with the same **natural frequency** of vibration (*resonant frequency*) as that of the vibrating tuning fork. The vibrations of the tuning fork are exciting the molecules that comprise the column of air within the length of tube above the water line. Amplitude becomes maximal (i.e., resonance

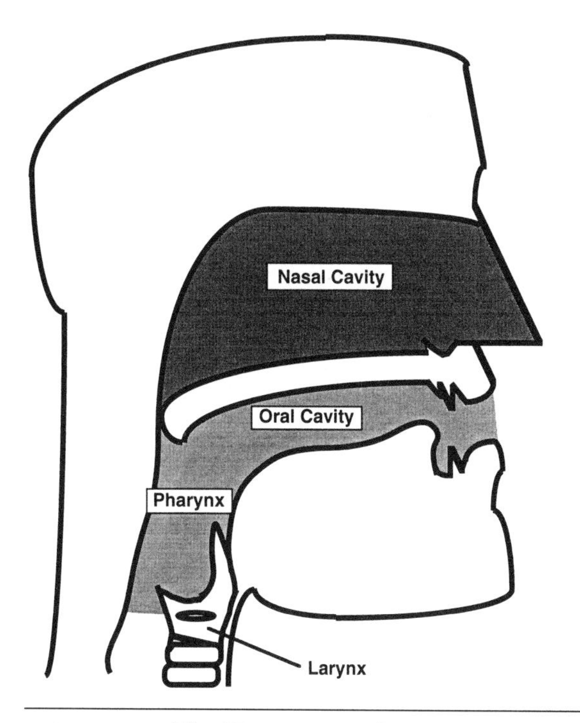

FIGURE 4-4 The Human Vocal Tract

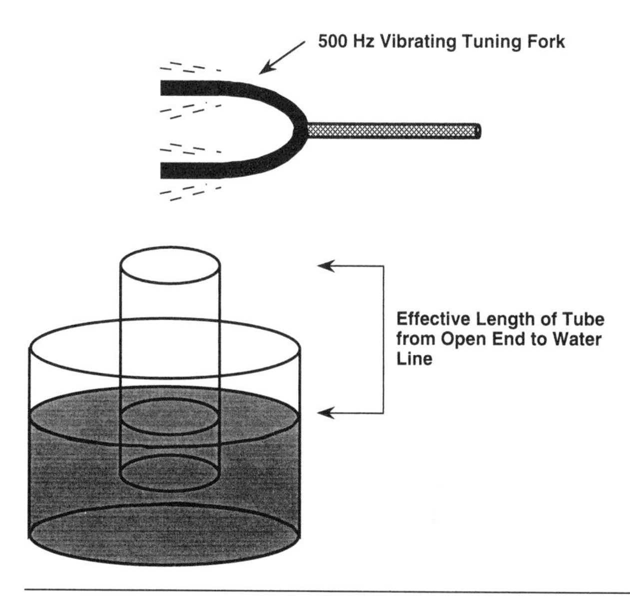

FIGURE 4-5 Laboratory Demonstration of Resonance

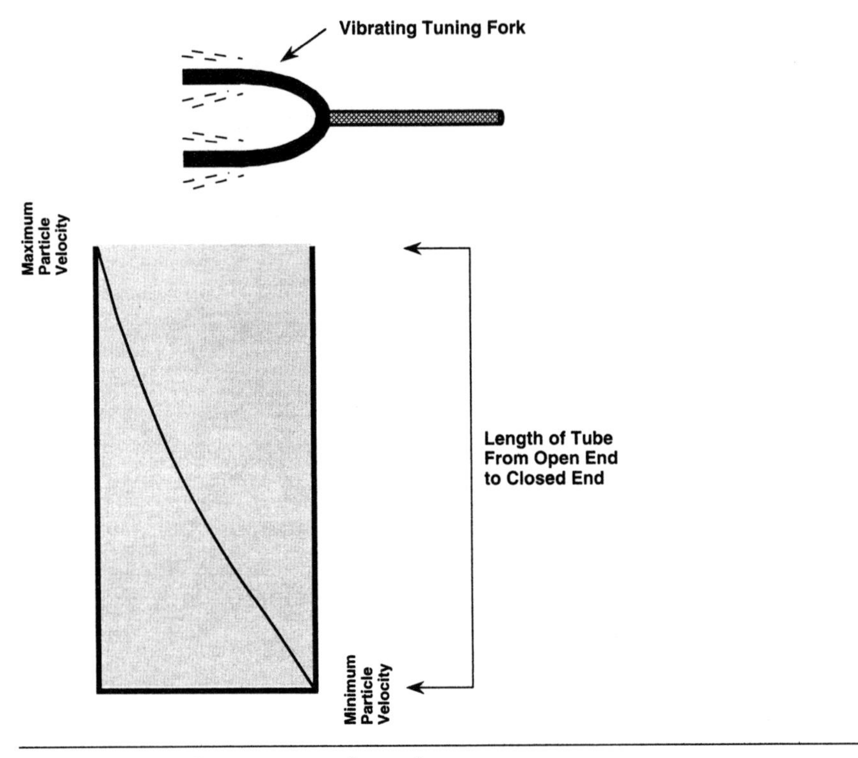

FIGURE 4-6 Velocity Curve for Tube Resonance

occurs) when the tube length is such that standing wave patterns representing air molecular velocity (speed) and pressure (force per unit area) are established within the tube (Figures 4-6 and 4-7).

The standing wave pattern referring to the velocity (speed) of the air molecules within the tube represents a condition in which the molecules are showing maximum velocity at the open end of the tube and minimum velocity at the closed end of the tube when the tube is being excited by the vibrations of the tuning fork. The speed of movement of the molecules between the extreme ends of the tube follow a curvilinear line, which is represented in Figure 4-6. The velocity or speed of air particle movement is in reference to the oscillatory movements of air molecules around a fixed point. In Chapter 3 an explanation was given to explain that air molecules, because of the elastic and inertial properties of the medium, do not move very far from a fixed point (*rest position*) when they are set in motion. Sound, which is a disturbance in the particles of a medium, travels through the medium, but the medium's particles remain relatively

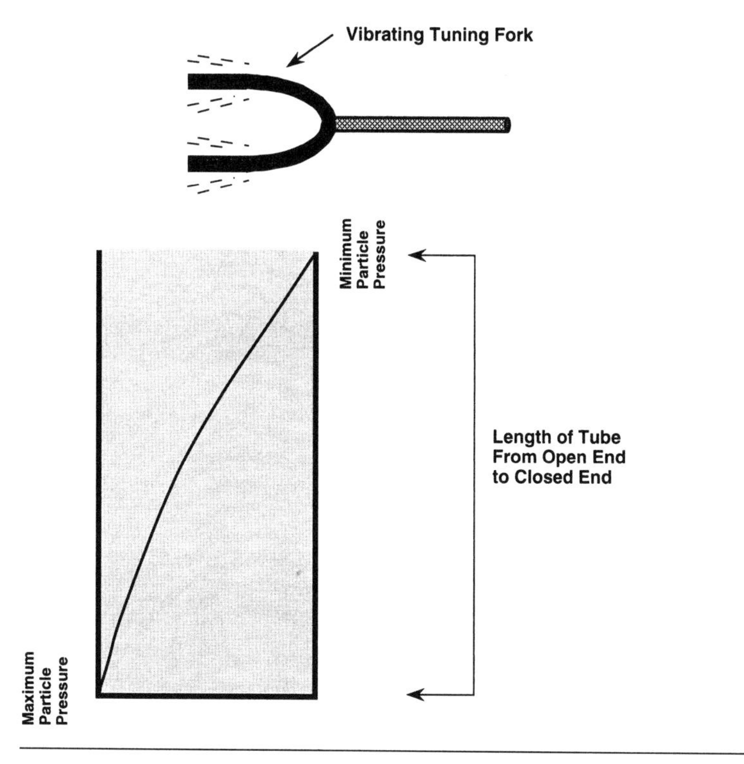

FIGURE 4-7 Pressure Curve for Tube Resonance

fixed, allowing the sound disturbance to cause them to oscillate about a fixed point to which they are anchored. Therefore, the speed of the oscillations of the air particles would be greatest at the open end of the tube and least at the closed end of the tube when the tube is excited by a sound source having the same natural frequency of vibration (resonant frequency) as the tube itself. Tube length, as suggested by the laboratory demonstration displayed in Figure 4-5, appears to be a critical factor in the determination of the natural, resonant frequencies at which the column of air inside a particular tube will vibrate.

A pressure curve (force per unit area) can also be drawn to represent molecular activity within the tube when resonance is occurring (see Figure 4-7). The pressure curve shows that while resonance is occurring, there is minimum particle pressure at the open end of the tube and maximum particle pressure at the closed end of the tube. The pressures on the molecules between the extreme ends follow a curvilinear line, as shown in Figure 4-7.

There appears to be an inverse relationship between the velocity and pressure curves that represent cavity or tube resonance (see Figures 4-6 and 4-7). The speed of oscillatory behavior for a particular air molecule is greatest at points along the length of the tube where the pressure being applied to that molecule by the vibrating sound source is least (i.e., the open end of the tube). Conversely, the speed of oscillatory behavior for a particular air molecule is least at points along the length of the tube where the pressure being applied to that molecule by the vibrating sound source is greatest (i.e., the closed end of the tube).

If the tuning fork in the laboratory demonstration described in Figure 4-5 had a natural frequency of 500 Hz, and was set into vibration over the open end of the tube, careful movement of the tube up and down in the water would show that resonance would occur in the tube when its length from its open end to the water line is 17 centimeters (cm) (Figure 4-8). This length of tube would allow for the establishment of the velocity and pressure standing wave patterns needed for resonance to occur within the tube when the tube is being excited by a 500-Hz sound source (the tuning fork). The speed of molecular movement and the pressure on the air molecules making up the column of air in the tube would be such that the appropriate standing wave patterns would be established (see Figures 4-6 and 4-7). It should be noted that tube length, which has been adjusted to accommodate the excitatory frequency of the tuning fork (500 Hz), is critical to the resonance characteristics of the tube, because tube length needed for resonance to occur is equal to the wavelength (λ) of the frequency of the sound source (the 500-Hz tuning fork) stimulating the tube, divided by a factor of 4, as indicated below:

$$\textit{Tube Length} = \frac{\textit{Wavelength } (\lambda)}{4}$$

To determine tube length needed for resonance to occur when it is being excited by a 500-Hz tuning fork, it is first necessary to determine the wavelength of that particular excitatory frequency. Wavelength (λ) is equal to the speed of sound in air (approximately 340 meters per second) divided by the stimulus frequency, as shown below:

$$\textit{Wavelength } (\lambda) = \frac{\textit{Velocity of Sound in Air}}{\textit{Stimulus Frequency}}$$

In this specific example, the wavelength for a 500-Hz signal would be 340 meters (or 34,000 cm) per second (velocity of sound in air)

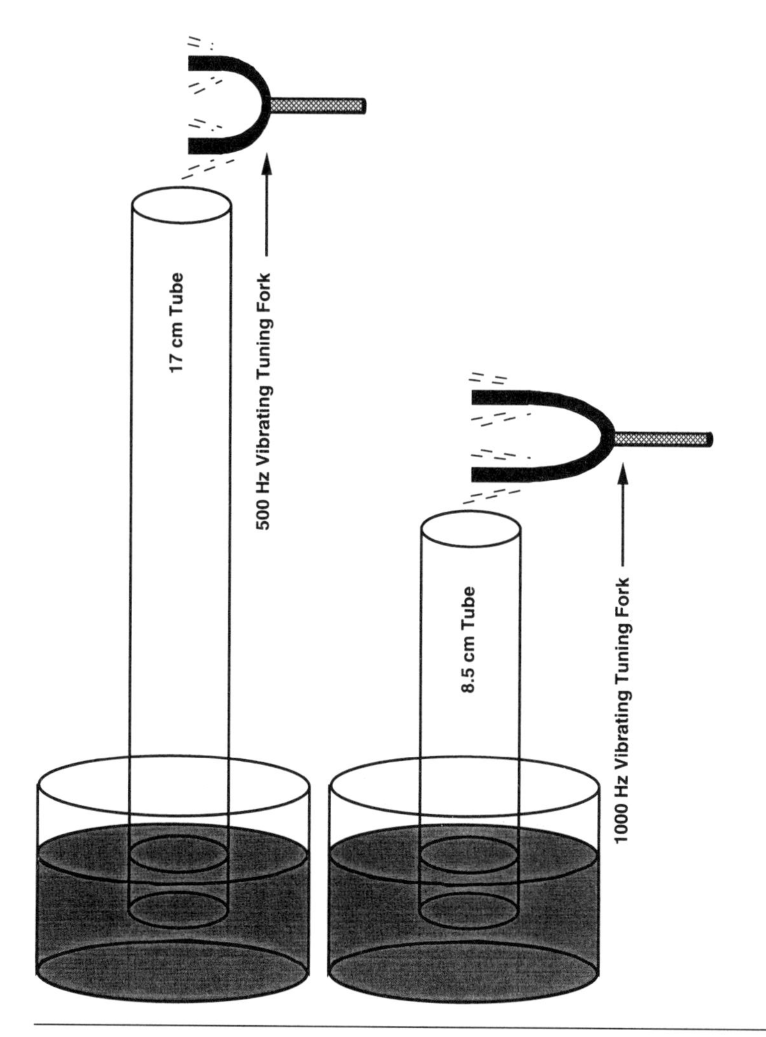

FIGURE 4-8 Tubes of Different Lengths

divided by 500 Hz (stimulus frequency), which is equal to 0.68 meters (or 68 centimeters):

$$Wavelength = \frac{340\ m/sec}{500\ Hz} = 0.68\ m = 68\ cm$$

Since the tube length needed for resonance to occur is equal to wavelength divided by a factor of 4, a wavelength of 68 cm divided by 4 is equal to a tube length of 17 cm, which is the tube length required for cavity (tube) resonance to occur when the tube is being activated by a tuning fork with a natural frequency of vibration of 500 Hz:

$$\text{Tube Length} = \frac{68\ cm}{4} = 17\ cm$$

If the tuning fork had a natural frequency of vibration of 1,000 Hz, the tube length needed for resonance to occur would be 8.5 cm (see Figure 4-8). Wavelength, in this example, would be equal to 340 meters per second (velocity of sound in air) divided by 1,000 Hz (the stimulus frequency), which would be 0.34 meters or 34 centimeters:

$$\text{Wavelength} = \frac{340\ m/sec}{1{,}000\ Hz} = 0.34\ m = 34\ cm$$

Thus, the tube length needed for resonance to occur is equal to wavelength divided by a factor of 4, which is 8.5 cm when the tube is being excited by a tuning fork with a natural frequency of 1,000 Hz:

$$\text{Tube Length} = \frac{34\ cm}{4} = 8.5\ cm$$

It would appear from these two examples that the shorter (8.5 cm) tube demonstrated a higher resonant frequency (natural frequency of vibration) than the longer (17 cm) tube. As the length of a tube closed at one end and open at the other (with uniform cross-sectional dimensions throughout its length) is either increased or decreased, its resonance characteristics (i.e., its natural, resonant frequencies) are altered in an orderly pattern: As tube length is increased, the natural (resonant) frequencies of vibration for the tube become lower and, conversely, as tube length is decreased, the natural (resonant) frequencies of vibration for the tube become higher.

For example:

$$\text{Tube Length} = 15\ cm$$

$$\text{Wavelength} = 4 \times 15 = 60\ cm$$

$$f_1 = \frac{\text{Velocity}}{\text{Wavelength}} = \frac{34{,}000\ cm/sec}{60\ cm}$$

$$f_1 = 567\ Hz$$

Tube Length = 17 cm

$$Wavelength = 4/1 \times 17 = 68\ cm$$

$$f_1 = \frac{Velocity}{Wavelength} = \frac{34{,}000\ cm/sec}{68\ cm}$$

$$f_1 = 500\ Hz$$

If tube systems are stimulated by an energy source that contains more than one natural frequency of vibration, then multiple resonances will occur simultaneously. In such cases, standing wave patterns are established for each resonant frequency occurring within the tube. The standing wave patterns for the lowest resonant frequency in terms of molecular (particle) velocity and pressure will be the same as those described in the tuning fork examples (see Figures 4-6 and 4-7).

There is a systematic relationship between the resonant frequencies in the tube model being discussed. A straight tube (one that is uniform in cross-sectional dimensions throughout its length) closed at one end and open at the other end can be multiply resonant when excited by a sound source containing more than a single natural frequency of vibration (see Figure 4-8). The first (lowest) resonant frequency for this tube of uniform cross-sectional dimensions throughout its length is equal to the frequency of a sound wave whose wavelength (λ) is four times the length of the tube:

$$f_1 = \frac{Velocity}{Wavelength\ (4 \times Tube\ Length)}$$

The tube's other (higher) resonant frequencies are odd-numbered multiples ($2n - 1$) of the lowest resonant frequency. Thus, its second resonant frequency is equal to the frequency of a sound wave whose wavelength is 4/3 times the length of the tube:

$$f_2 = \frac{Velocity}{Wavelength\ (4/3 \times Tube\ Length)}$$

The third resonant frequency for this tube is equal to the frequency of a sound wave whose wavelength is 4/5 times the length of the tube:

$$f_3 = \frac{Velocity}{Wavelength\ (4/5 \times Tube\ Length)}$$

For example, for a 17-cm tube open at one end and closed at the other end, of uniform cross-sectional dimensions throughout its length (i.e., constant shape):

$$f_1 = \frac{34{,}000\ cm/sec}{4/1 \times 17\ cm} = \frac{34{,}000\ cm/sec}{68\ cm} = 500\ Hz$$

$$f_2 = \frac{34{,}000\ cm/sec}{4/3 \times 17\ cm} = \frac{34{,}000\ cm/sec}{22.67\ cm} = 1{,}500\ Hz$$

$$f_3 = \frac{34{,}000\ cm/sec}{4/5 \times 17\ cm} = \frac{34{,}000\ cm/sec}{13.6\ cm} = 2{,}500\ Hz$$

However, once the tube is altered so that it is not uniform in cross-sectional dimensions (i.e., a tube that is not straight), this specific relationship between the resonant frequencies no longer exists.

Thus, resonance is a characteristic of all periodic vibrating systems that enables them to respond strongly to oscillatory disturbances that are the same as their own natural frequencies of vibration. Vibrating systems are capable of ignoring oscillatory disturbances at frequencies that do not match their own natural frequencies of vibration. If a system is excited at its natural (resonant) frequencies (those to which it responds most vio-

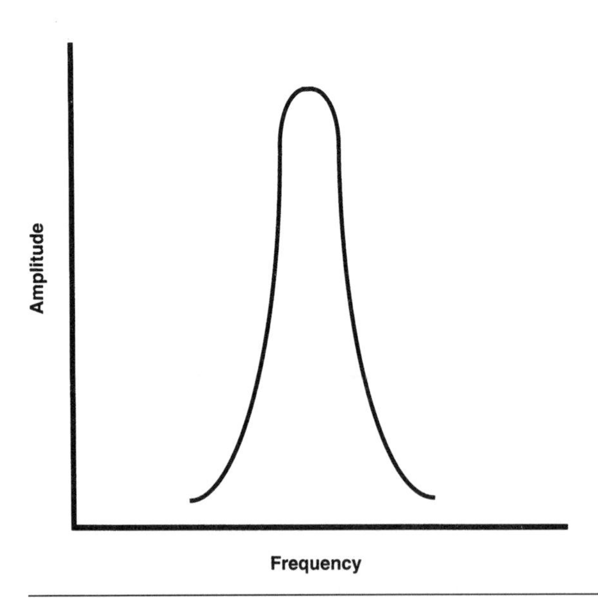

FIGURE 4-9 Sharply Peaked Frequency Response Curve of an Undamped Resonator

lently), then it is resonating, and the result is an increase in overall amplitude (energy).

Frequency Response Curve

Cavities and tubes can serve as resonators because they contain a column of air capable of vibrating at certain frequencies (their resonant or natural frequencies). A graph of the frequencies to which a resonator will respond (resonate) can be constructed; this graph is called a **frequency response curve** or **resonance curve.**

Undamped resonators are those that resonate to a narrow range of frequencies, because they themselves contain only a narrow range of frequencies. Figure 4-9 displays the sharply peaked frequency response curve of an undamped resonator, a tuning fork (which vibrates at only one frequency). Damped resonators, on the other hand, resonate to a broad range of frequencies and, as shown in Figure 4-10, are characterized by a flat, broad frequency response curve. The range of frequencies to which a resonator responds (i.e., the range of a resonator's natural or resonant frequencies) is called its **bandwidth.** The frequency response curve of the resonator displayed in Figure 4-10 shows a broad bandwidth.

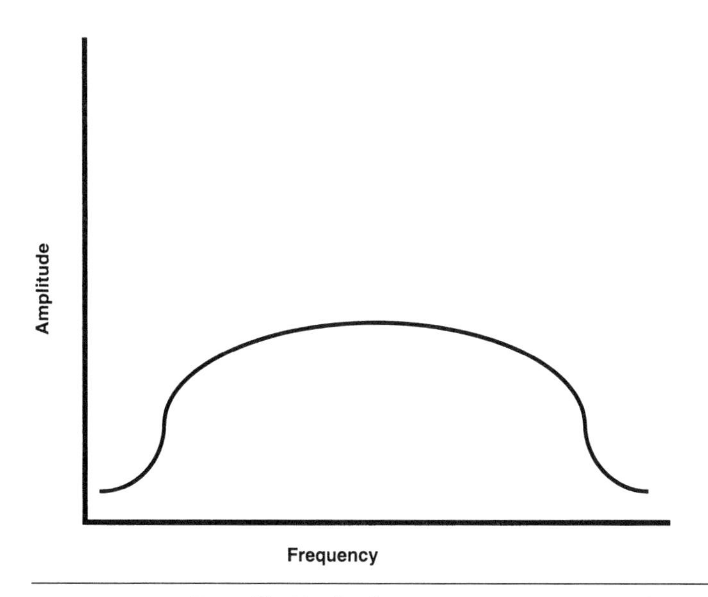

FIGURE 4-10 Broadly Peaked Frequency Response Curve of a Damped Resonator

Vocal Tract Analogy

The foregoing tube model is of interest to speech scientists because of its analogy to the human vocal tract system (Figure 4-11). The human vocal tract is comprised of orifices, cavities that extend between the larynx and the lips. More specifically, the human vocal tract consists of the pharynx, oral cavity, nasal cavity, and the lips, which can add an extension to the anterior portion of the oral cavity beyond the teeth. Thus, the human vocal tract system includes the major articulators: tongue, teeth, lips, hard palate, soft palate (velum), and pharynx.

The human vocal tract system is analogous to a tube closed at one end (the larynx, when the vocal folds are adducted for the production of voiced speech sounds), open at the other end (the lips), and uniform in cross-sectional dimensions throughout its length [when producing a neutral vowel sound, like /ə/, (as in **a**bout)]. Excitation of the vocal *tract* can be accomplished by vocal *fold* vibrations within the larynx (Figure 4-12). The larynx provides complex vibrations containing more than one frequency, and it stimulates the vocal tract from the tract's "closed end." The glottal slit between the vibrating vocal folds is so much smaller than the slit between the lips that, in a mathematical sense, the end of the vocal tract, which begins at the glottis, can be viewed as closed.

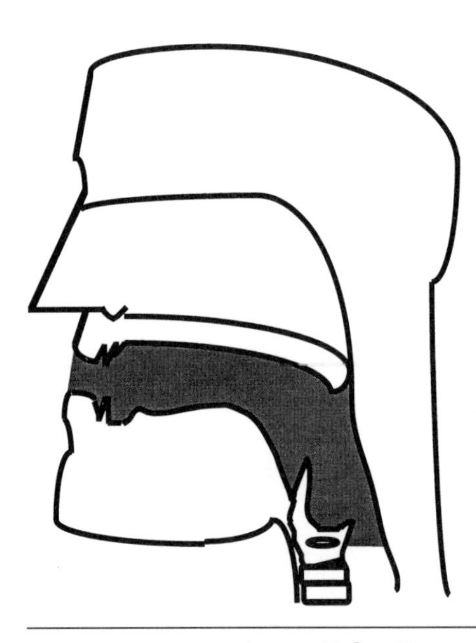

FIGURE 4-11 Inanimate Tube–Human Vocal Tract Analogy

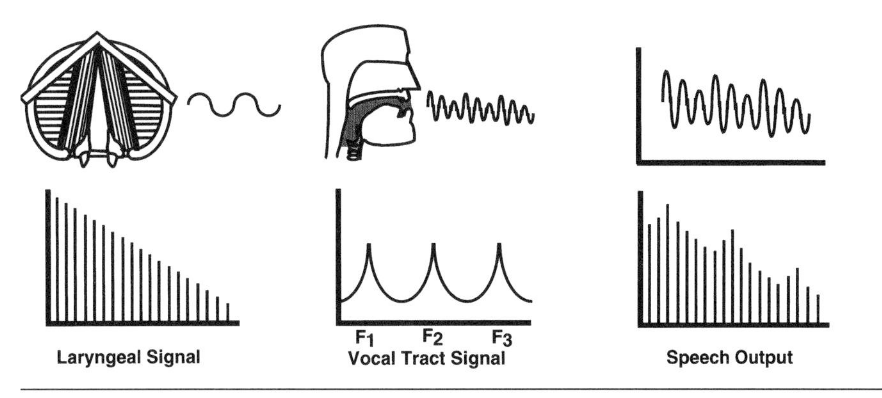

FIGURE 4-12 Effects of Acoustic Filtering

The vocal tract of the average adult male is approximately 17 cm in length when measured from the vocal folds to the lips (Figure 4-13). If this 17-cm vocal tract is shaped for the production of the schwa vowel /ə/ (as in **a**bout), it is analogous to a tube system closed at one end, open at the other end, and uniform in cross-sectional dimensions throughout its length. When excited by the complex, quasiperiodic, multifrequency sound source being generated at the larynx, this vocal tract shape will allow resonances within the tube to occur at 500 Hz, 1,500 Hz, and 2,500 Hz. The vowel sound heard will be the schwa vowel /ə/ (Figure 4-14).

The length of the vocal tract is an important factor in the phenomenon of resonance. As was the case with the inanimate tube system described earlier, alteration of vocal tract length will affect its natural, res-

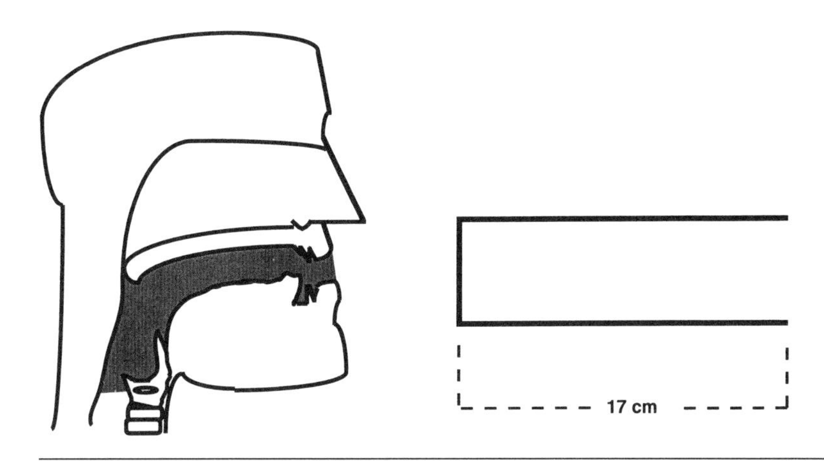

FIGURE 4-13 Inanimate Tube Analogy of Adult Male Vocal Tract

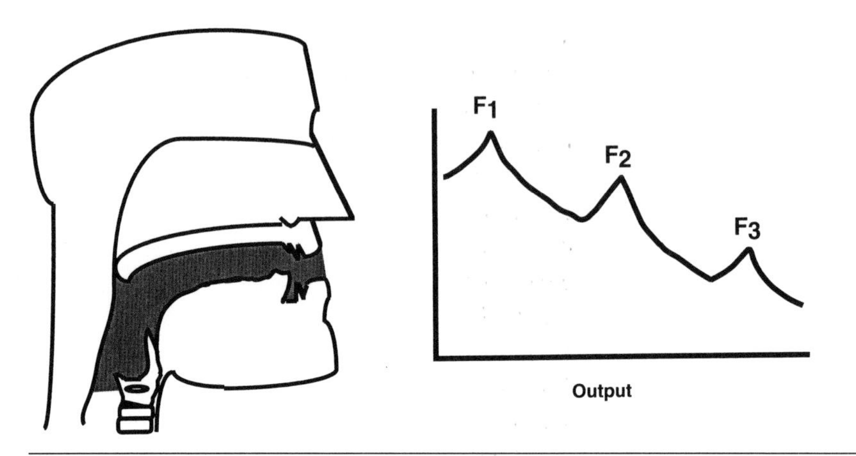

FIGURE 4-14 Vocal Tract Output for Schwa Vowel

onant frequencies. On average, adult males have longer vocal tracts than do adult females, and adult females, in turn, have longer vocal tracts than do children. Therefore, it would be expected that production of a given vowel by adult female speakers would be accomplished with higher resonant frequencies than those for adult males, and, similarly, the production of the same vowel by children would involve higher resonant frequencies than those for adult females. In each case, the same vowel sound would be heard, but the resonant frequencies of the respective

TABLE 4-1 *Average Formant Frequences (Hz) of Vowels by 76 Speakers*

		i	ɪ	ɛ	æ	ɑ	ɔ	ʊ	u	ʌ	ɝ
F_1	Adult male	270	390	530	660	730	570	440	300	640	490
	Adult female	310	430	610	860	850	590	470	370	760	500
	Child	370	530	690	1010	1030	680	560	430	850	560
F_2	Adult male	2290	1990	1840	1720	1090	840	1020	870	1190	1350
	Adult female	2790	2480	2330	2050	1220	920	1160	950	1400	1640
	Child	3200	2730	2610	2320	1370	1060	1410	1170	1590	1820
F_3	Adult male	3010	2550	2480	2410	2440	2410	2240	2240	2390	1690
	Adult female	3310	3070	2990	2850	2810	2710	2680	2670	2780	1960
	Child	3730	3600	3570	3320	3170	3180	3310	3260	3360	2160

(Modifed with permission from G.E. Peterson & H.L. Barney, Control methods used in a study of vowels. *Journal of the Acoustical Society of America 24,* 1952, 183.)

vocal tracts would be different, depending on their overall length (Table 4-1). Therefore, for every frequency, there is a certain length of a resonator that will produce maximum resonance; the lower the frequency, the longer the tube or cavity must be; and the higher the frequency, the shorter the tube or cavity.

Once the shape of the vocal tract is altered to produce any of the speech sounds of English other than the schwa vowel, the relationship between the resonances noted for the straight inanimate tube analogy no longer holds true. Thus, the human vocal tract is analogous to a tube closed at one end, open at the other end, and uniform in cross-sectional dimensions, but only when it is approximating a configuration required for the production of the neutral schwa vowel. However, once the vocal tract shape is altered, as it would be for the production of other vowel sounds, the analogies and examples provided in this chapter require modification (Figures 4-15 and 4-16).

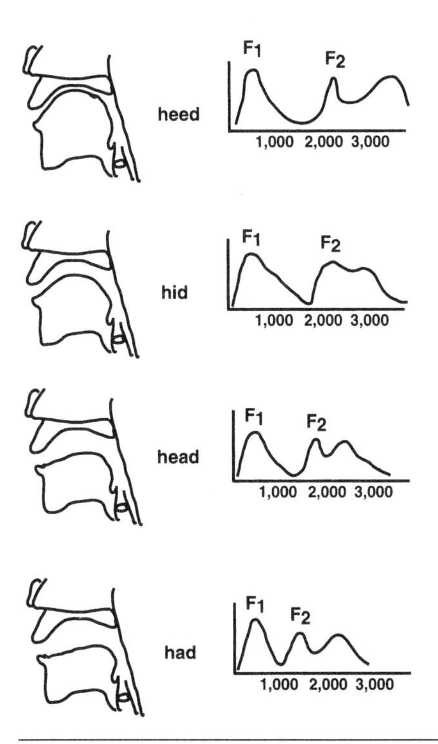

FIGURE 4-15 Vocal Tract Shapes and Corresponding Spectra for Front Vowels

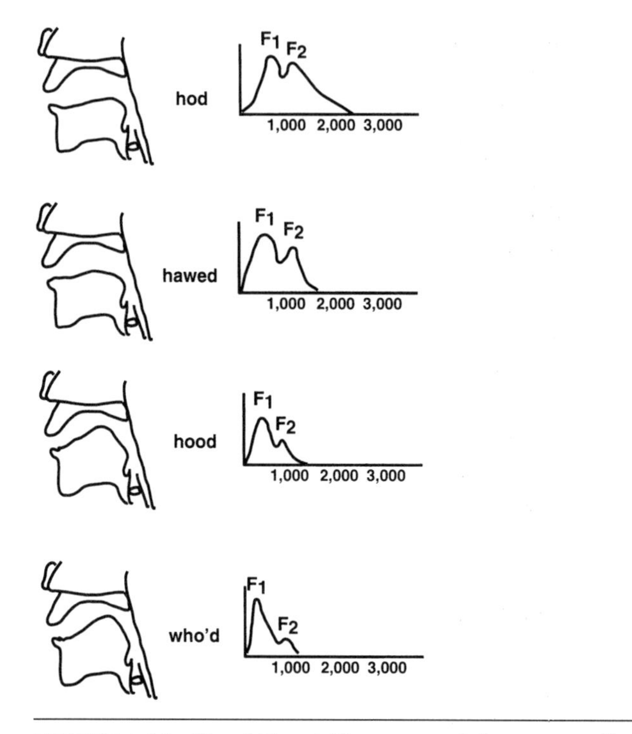

FIGURE 4-16 Vocal Tract Shapes and Corresponding Spectra for Back Vowels

Summary

The concept of resonance, particularly cavity (acoustical) resonance, is a very important one in understanding the articulatory aspects of the speech production process. The human vocal tract is analogous to a tube open at one end (the lips) and closed at the other end (vocal folds) and therefore parallels the systematic acoustic behavior associated with tubes or cavities. Thus, the phenomenon of resonance and its applications relate to the acoustics of speech production, which is discussed in detail in the next chapter.

Study Questions

1. Describe the application of the principle of resonance in the following:
   a. sympathetic vibration
   b. sounding-board effect
   c. cavity (acoustical) resonance
2. For a tube of uniform cross-sectional dimensions that is open at one end and closed at the other end,
   a. the *first resonant frequency* is equal to the frequency of a sound wave whose wavelength is _____ times the length of the tube.
   b. the *fifth resonant frequency* is equal to the frequency of a sound wave whose wavelength is _____ times the length of the tube.
3. What is the *eighth resonant frequency* of a 25-cm tube of uniform cross-sectional dimensions that is open at one end and closed at the other end?
4. What is the *wavelength* of the twentieth resonant frequency of a 10-meter tube of uniform cross-sectional dimensions that is open at one end and closed at the other end?
5. Define and illustrate the concept of *frequency response curve.*

Suggested Readings

Denes, P., & Pinson, E. (1972). *The Speech Chain.* New York: Bell Telephone Laboratories.

Kent, R.D., & Real, C. (1992). *The Acoustic Analysis of Speech.* San Diego: Singular Publishing Group.

Lieberman, P., & Blumstein, S. (1988). *Speech Physiology, Speech Perception, and Acoustic Phonetics.* New York: Cambridge University Press.

Speaks, C.E. (1996). *Introduction to Sound: Acoustics for the Hearing and Speech Sciences.* (Second Edition). San Diego: Singular Publishing Group.

5

Acoustics of Speech Production

The study of speech production from an acoustical point of view provides the means for looking at a very complex process in a simpler way. The body tissues used in speech production and the ways in which they operate to produce the speech signal represent an amazingly complex activity that provides a primary channel for human communication. Moreover, it is difficult to view this complex process in action because of the location and interrelationships of the tissues involved. They are hidden from view and connected to the nervous system in such a way that it is impossible to study them directly. However, the acoustical signal produced by the intricate movements of muscles and bones that comprise the speech production mechanism can be recorded and studied directly. Through careful analyses of the acoustical signal, inferences can be made as to how the actual speech production process takes place.

Acoustical Model of Speech Production

The acoustical model (also called the *source-filter theory*) of speech production will serve as the basis for the information presented in this chapter (Figure 5-1). This model has been constructed through the careful study of the acoustical signal generated during speech production. It per-

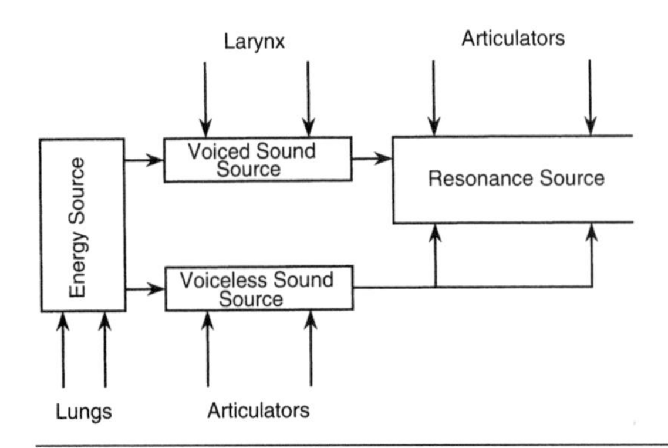

FIGURE 5-1 Acoustical Model of Speech Production

mits the breakdown of the speech production process into three major components:

1. Energy source (ES)
2. Sound sources (SS)
3. Resonance source (RS)

These three components combined represent a **generative** (or **modulatory**) **system** of speech production. Each component is modified by the one that succeeds it, leading to an acoustical output that travels through a medium from the speaker to the listener. It is assumed that there is a central controlling mechanism, the **central nervous system,** for the three parts of this model. The central nervous system contains all of the necessary elements for the encoding (creation) of messages to be spoken, incorporating a language tool for this purpose.

Component #1: Energy Source

The energy source for speech production is supplied by the lungs. It is composed of air that is exhaled from the lungs and controlled in such a manner as to provide a reliable energy supply for the other components of the speech production mechanism (Figure 5-2). This energy supply is referred to as *direct current (dc) air flow* because it is very steady during speech production; the respiratory structures of the thorax and abdomen perform complex maneuvers to keep it that way. This steady flow of exhaled air provides a constant air pressure that is delivered to the sound

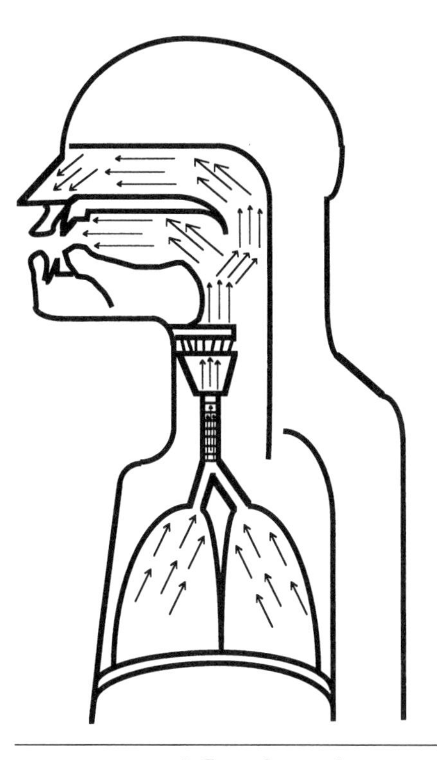

FIGURE 5-2 Airflow from the Lungs

sources of the speech production mechanism, where it is converted into audible sound.

Component #2: Sound Sources

As shown in Figure 5-1, the speech production mechanism contains two major sound sources for converting exhaled air into audible sound:

1. The voiced sound source, located in the larynx, provides periodic modulation of the air supply, which is tonal in quality. This type of sound is important for the production of vowels and voiced consonants (Figure 5-3).

2. The voiceless sound source, located in various parts of the articulatory system, provides aperiodic modulation of the air supply, which is noise-like in quality (Figure 5-4). The aperiodic modulation of the air

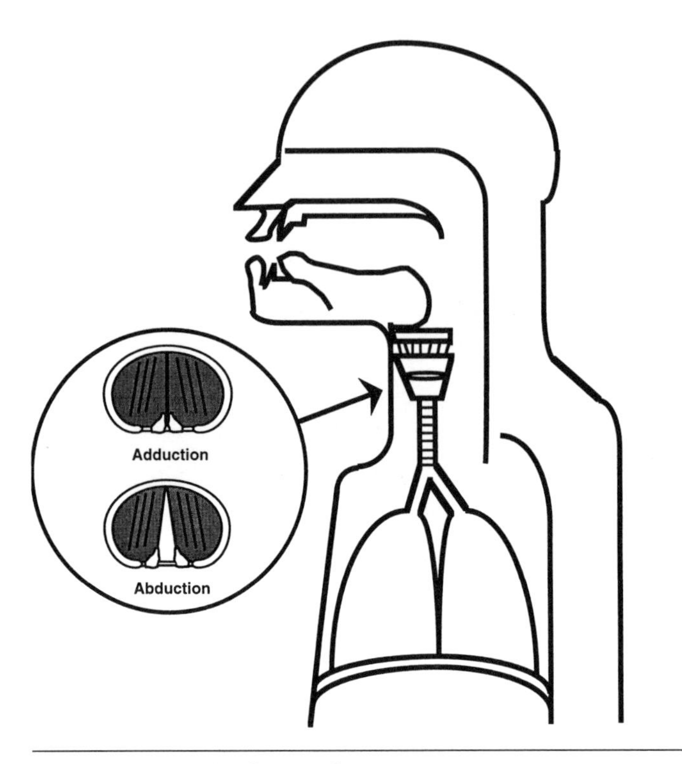

FIGURE 5-3 Voiced Sound Source

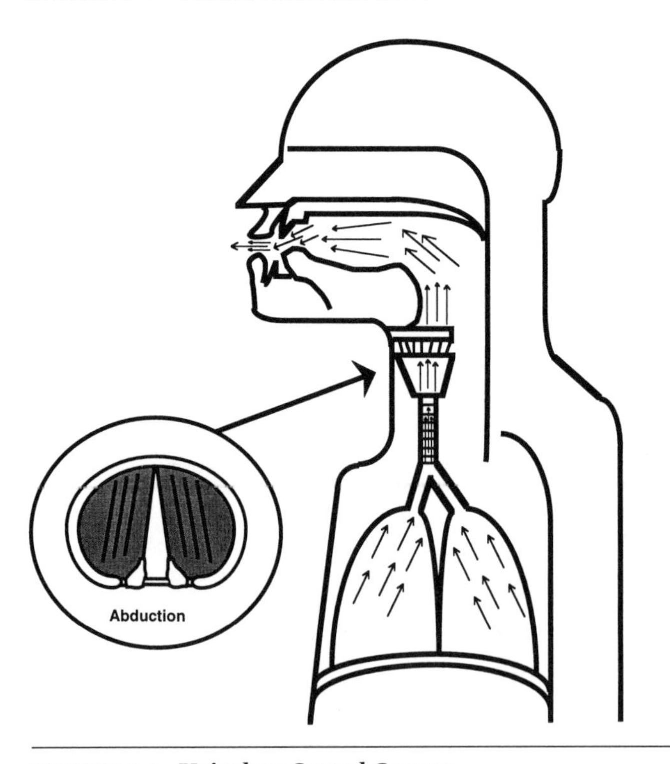

FIGURE 5-4 Voiceless Sound Source

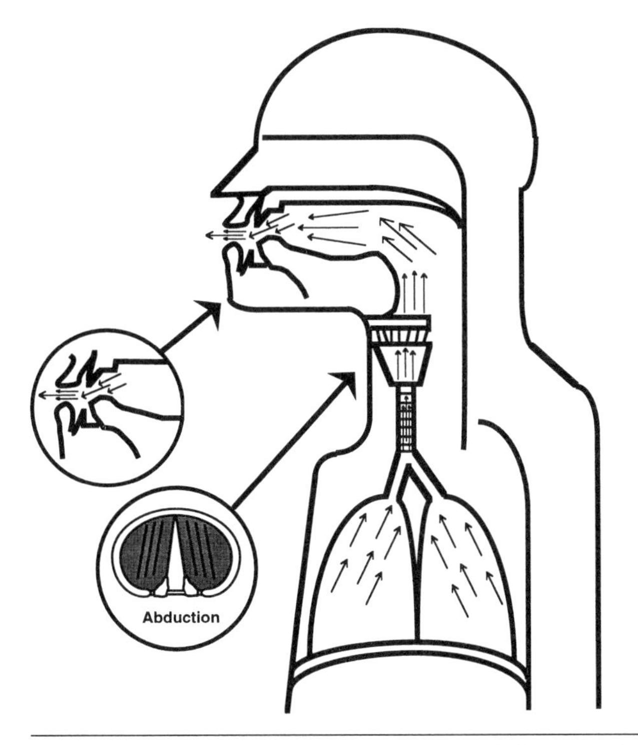

FIGURE 5-5 Continuous Aperiodic Modulation of Air Supply Provided by Voiceless Sound Source for Fricatives

supply provided by the voiceless sound source can be continuous for the production of certain speech sounds, such as fricatives (Figure 5-5), or it can be preceded by silence for the production of other speech sounds, such as stop plosives (Figure 5-6).

The two sound sources essentially provide three types of possible modulation (regulation, adjustment) of the direct current air supply being exhaled from the lungs: vibratory modulation (voiced sound source), turbulent modulation (voiceless sound source), and stepwise modulation (voiceless sound source preceded by silence).

Vibratory Modulation

Vibratory modulation is the result of the voiced sound source; it is important in the production of all vowels and voiced consonants. Vibratory modulation of the energy source (exhaled air supply) results in a quasiperiodic sound that has certain unique acoustical characteristics. This type of sound production, when analyzed spectrally, will provide a

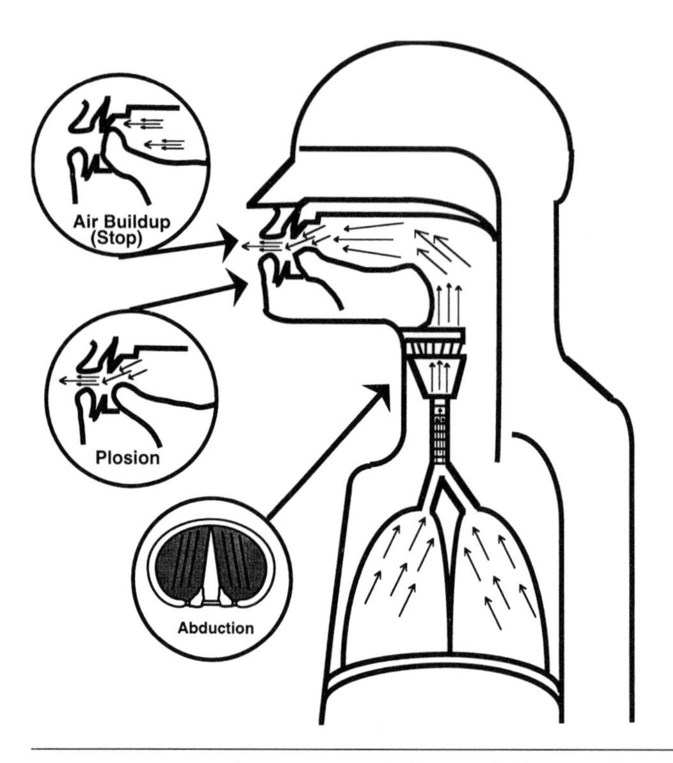

FIGURE 5-6 Continuous Aperiodic Modulation of Air Supply Provided by Voiceless Sound Source for Stop Plosives

line (discrete) sound spectrum (Figure 5-7A). The harmonics (those component pure tones above the fundamental) are whole-number multiples of the fundamental frequency (which is the lowest and most powerful frequency in the spectrum). The spectral display for vibratory modulation will also show an amplitude drop of 12 dB per octave with an increase in frequency. When vibratory modulation is analyzed as a waveform, the result is a sawtoothed pattern, which represents the openings and closings of the vocal folds as well as their degree of excursion from the midline during each cycle of vibration (Figure 5-7B). During phonation, the initial phase involves closure of the vocal folds, which is represented as the baseline on the waveform. The vocal folds begin to open as the subglottic air pressure underneath blows them apart until they reach an excursion point from midline where their own stiffness begins to bring them back toward the midline and closure again (Figure 5-8). This process repeats itself over and over again, providing a repetitive sawtoothed pattern on the waveform. The repetitiveness of the pattern indi-

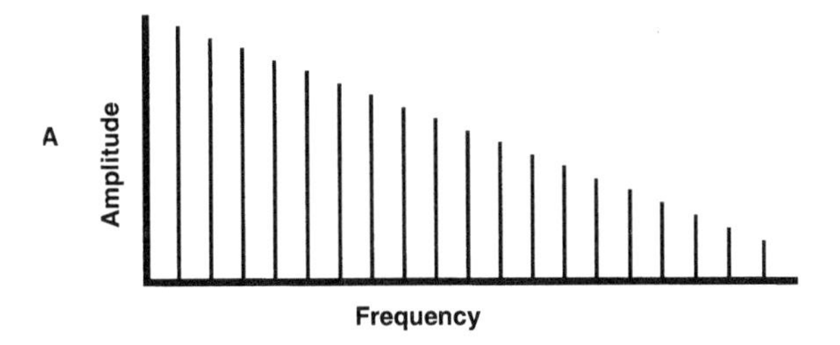

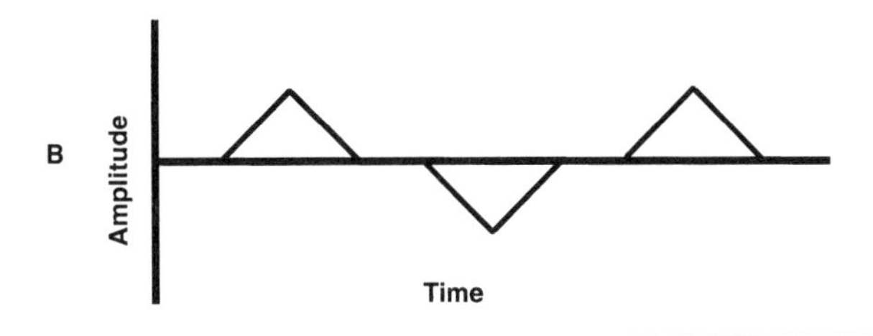

FIGURE 5-7 Spectrum (A) and Waveform (B) for Vibratory Modulation

cates that vibratory modulation is, by its very nature, periodic. This periodicity adds a tonal quality to the perceived signal.

Vocal Frequency. Fundamental frequency of the voice is an acoustical phenomenon directly related physiologically to the number of times per second that the vocal folds open and close (see Figure 5-8). The greater the number of openings and closings of the folds per unit time, the higher the fundamental frequency and perceived vocal pitch (the perceptual correlate of fundamental frequency) will be. Fundamental frequency is the lowest and most powerful frequency in the sound spectrum for vibratory modulation (see Figure 5-8). The average number of vibrations (openings and closings) of the male vocal folds is approximately 120 per second, which translates into an average vocal fundamental frequency of 120 Hz. For females, the number of vibrations of the vocal folds per second is approximately 220, making their average vocal fundamental frequency 220 Hz. In addition, the pitch of the female voice is, on average, perceived as being higher than that of the male voice because the vocal folds of the female usually vibrate more rapidly per unit time.

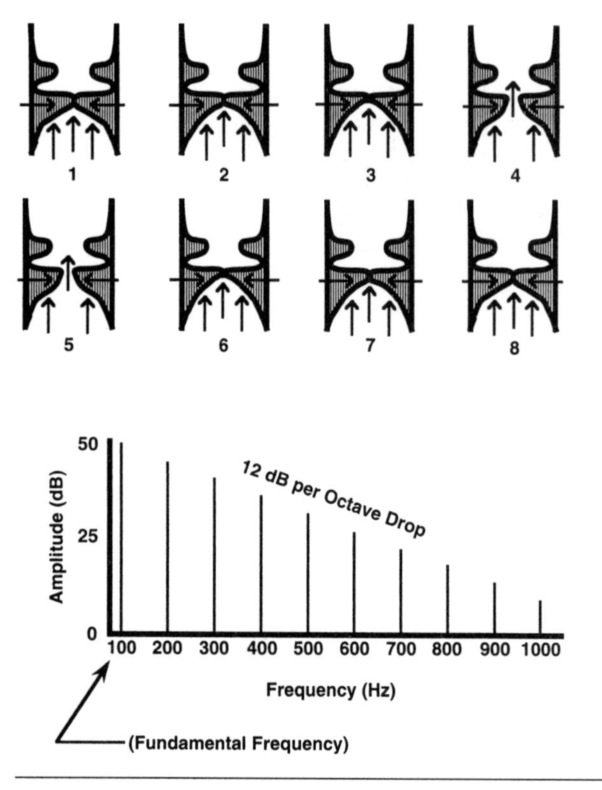

FIGURE 5-8 Vocal Frequency

Within certain physiological limits, an individual can alter his vocal fundamental frequency. One physiological explanation for this ability relates to the combined efforts of vocal fold stiffness (tension) and subglottal air pressure. As the vocal folds are stiffened, they undergo a change in mass (they become thinner), which will facilitate more rapid vibrations. Therefore, the number of vibrations per second is increased, provided the subglottic air pressure is increased to overcome the added resistance needed to burst the folds apart. The limits provided by the physiological capabilities of the larynx usually allow most individuals to alter their vocal fundamental frequency within a range of one octave (eight notes). Moreover, gender differences exist based on average physiological variations between females and males. Females usually have thinner vocal folds than males and therefore their folds vibrate more rapidly than do those of males.

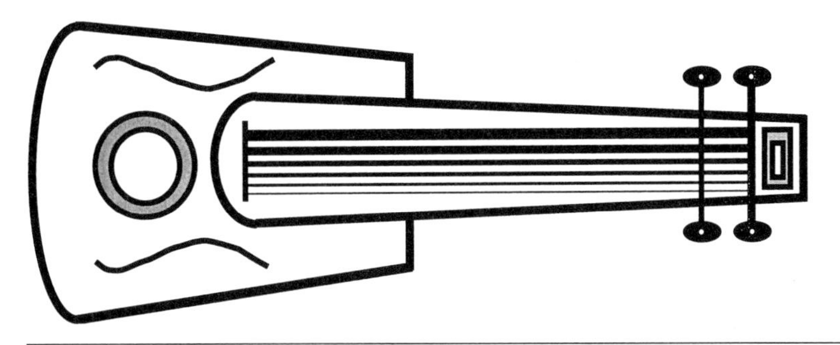

FIGURE 5-9 Stringed Instrument Showing Thick and Thin Strings

The analogy of the vocal folds to a stringed instrument is relevant in a discussion of gender differences in vocal fundamental frequency (Figure 5-9). If we observe the strings on a violin or guitar or other stringed instrument, we notice that they differ in thickness and, when struck, bowed, plucked, and so on, the thinner strings produce an audibly higher pitch than do the thicker strings. The same is true of the human vocal folds: Thinner folds produce higher rates of vibration (higher vocal fundamental frequency) and higher perceived pitch; thicker folds vibrate more slowly, yielding lower vocal fundamental frequency and, consequently, lower perceived pitch.

Vocal Intensity. Vocal intensity, like vocal fundamental frequency, is an acoustical (physical) phenomenon. It is directly related to the degree of opening created between the vocal folds during phonation (Figure 5-10). The farther apart the vocal folds are spread during each cycle of vibration, the greater is the speaker's vocal intensity. In general, vocal loudness, which is the perceptual counterpart of vocal intensity, will be perceived as greater with increases in vocal intensity. Vocal intensity is directly related to the amount of air pressure built up below the vocal folds (**subglottic air pressure**) (Figure 5-11). The greater the subglottic air pressure, the greater the vocal intensity and, consequently, the greater the perceived vocal loudness. Anyone who has ever been engaged in shouting realizes that it takes greater amounts of air supply for an increase in loudness to occur. As the speaker increases the perceived loudness of her voice, she is automatically increasing subglottic air pressure and, in turn, the vocal folds are spreading apart farther from the midline during each phonatory cycle.

The descriptions of how vocal frequency and vocal intensity are produced by the laryngeal mechanism are based on assumptions underlying

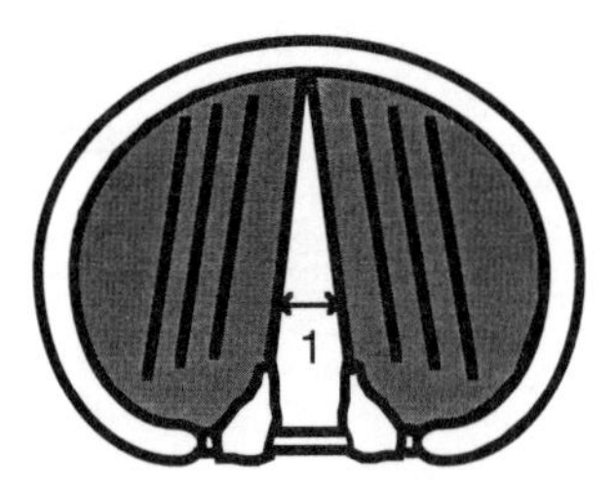

Excursion #1
(Less Vocal Intensity)

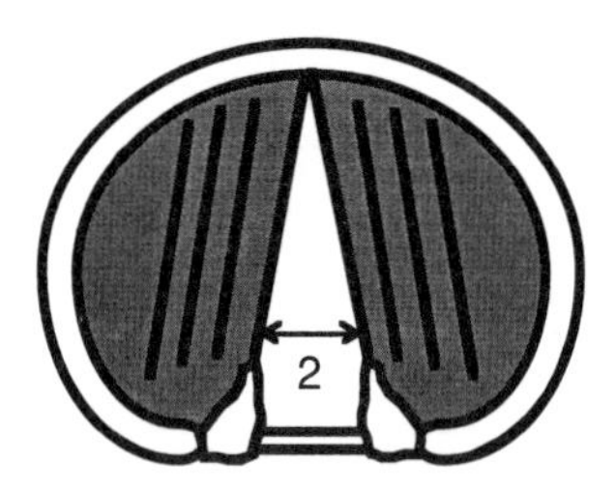

Excursion #2
(More Vocal Intensity)

FIGURE 5-10 Vocal Fold Excursions

the **myoelastic–aerodynamic theory of phonation,** the most popular contemporary theory to explain the mechanism for vocal fold vibration. This theory assumes that for vibration of the vocal folds to take place, there has to be muscle activity (myoelastic aspect) which will approximate the vocal folds into a midline position just prior to and during the vibratory activity. The major force involved in voice production (aerodynamic aspect) is the air being exhaled from the lungs. The steady airflow from the lungs creates subglottic pressure below the approximated vocal folds. When this pressure becomes great enough, it overpowers the muscle tension of the vocal folds and, in turn, spreads them apart in a burstlike manner (Figure 5-12). As the air that had been built up below the folds flows through the **glottis** (the space between the vocal folds), a negative pressure is created between the vocal folds. Because of this negative pressure, the vocal folds are literally sucked together in the midline position, with air pressure building up below the folds again, and the cycle begins again. This activity will occur in the average adult male approximately 120 times

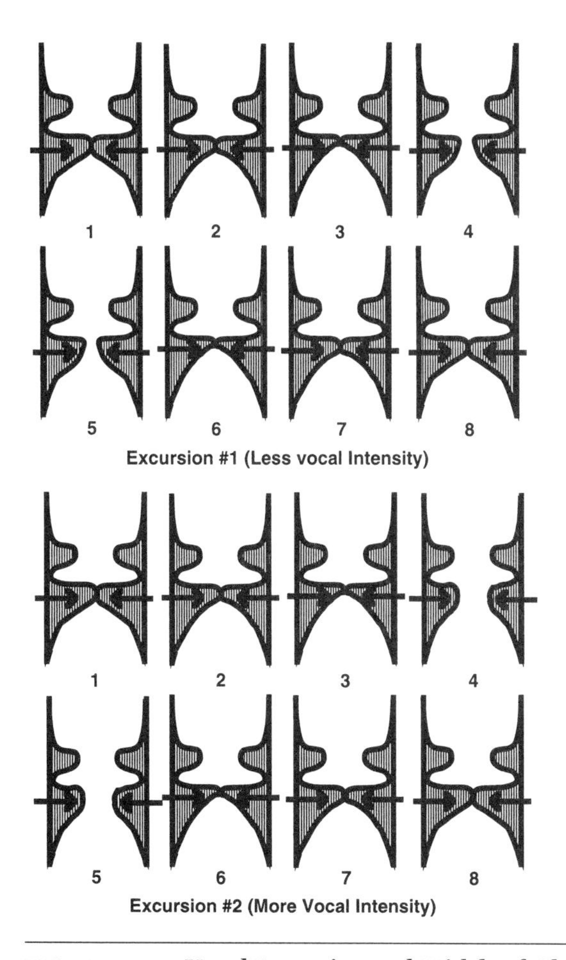

FIGURE 5-11 Vocal Intensity and Width of Glottis

per second and in the average adult female, approximately 220 times per second, producing the vocal fundamental frequency (and perceived vocal pitch) for the vibratory signal being generated.

The negative pressure that results between the vocal folds during vibratory modulation of the airstream is explained by **Bernoulli's Principle,** a basic principle of elementary physics. It states that, in tube systems, where the velocity (speed) of airflow is greatest, the pressure on the sides of the tube wall is least; thus, velocity and pressure are inversely related. In Figure 5-13A, the larynx is represented as a tube system that is constricted at the glottis so that the airflow through the relatively narrow glottis requires an increase in velocity to accommodate the larger volume of air below the constricted point (glottis). When the airflow

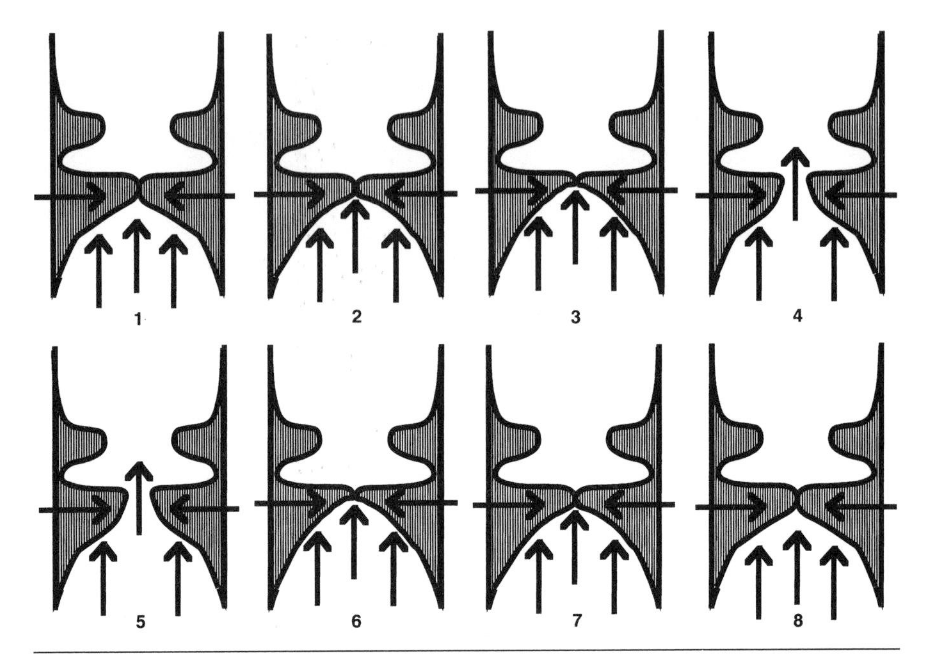

FIGURE 5-12 Subglottic Air Pressure and Vocal Intensity

increases in velocity as it passes through the glottis, a negative pressure is built up within the glottal space along the glottal margins, thus causing the folds to be sucked together in midline until another buildup of air beneath the folds (subglottic pressure) becomes sufficient to burst them open again (Figure 5-13B). Bernoulli's principle is operative for laryngeal vibratory modulation of the airstream because of the additional assumption that the intrinsic muscles of the larynx are effectively forcing the vocal folds into a midline position for phonatory purposes (myoelastic aspect). The myoelastic–aerodynamic theory of phonation is perhaps not so much a theory as it is a reasonable description of vibratory modulation of energy in the form of air which can be turned into pulses of audible sound for the purpose of speech production.

Turbulent Modulation

Turbulent modulation is the result of the voiceless sound source; it is important in the production of speech sounds that are classified as **fricatives** (also called **continuants**). These are continuous noise-like sounds such as /s/ (in **s**un). Turbulent modulation is produced by forcing exhaled air (energy source) through a constricted passageway that is created

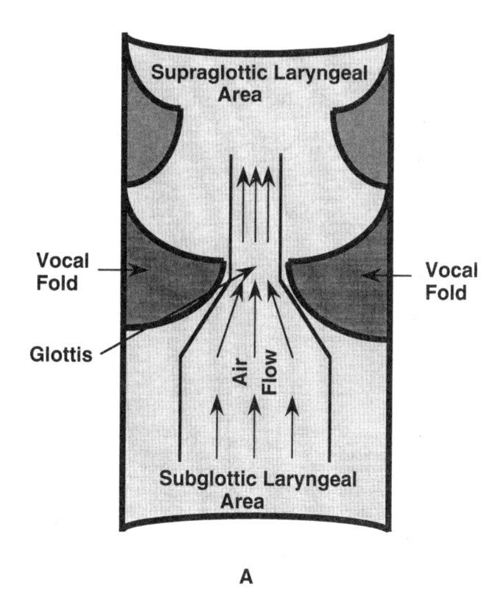

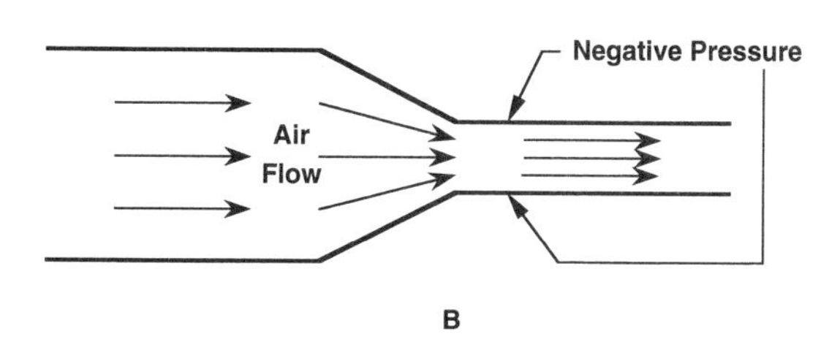

FIGURE 5-13 Bernoulli's Principle Applied to Laryngeal Vibratory Modulation of Airstream from Lungs

somewhere within the vocal tract. The vocal tract can be defined as a tube extending from the vocal folds of the larynx to the lips (Figure 5-14). It serves dual purposes in that it is capable of creating constrictions for turbulent and stepwise modulations of the exhaled airstream, while at the same time assuming different shapes for the production of the various speech sounds of a language (resonance). Thus, the vocal tract can be viewed as a sound generator (voiceless sound source) and, at the same time, as a "speech sound" producer (resonance). Turbulent modulation of the exhaled airstream results in a continuous aperiodic sound that is "noise-like" in quality. The spectrum for turbulent sound is similar to that of white noise, whereby there is an energy spread over the entire frequency range in fairly equal amounts. This energy spread is shown via the use of a spectral envelope (Figure 5-15A). Unlike the spectrum for

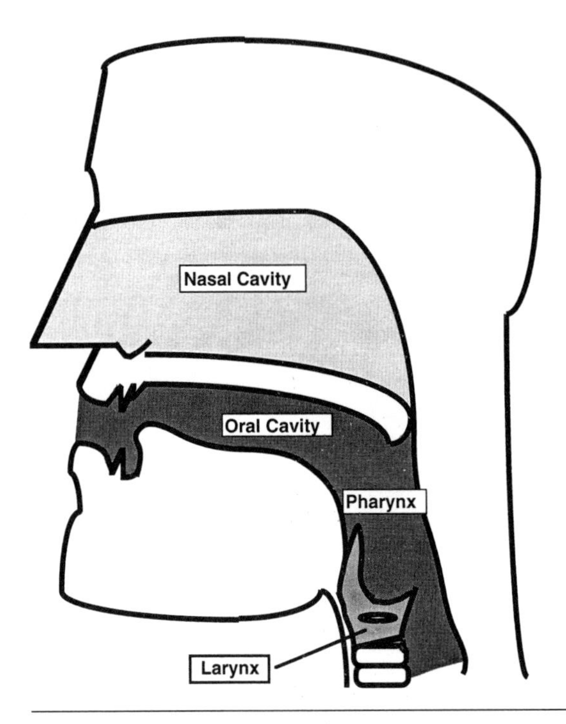

FIGURE 5-14 Vocal Tract

vibratory modulation, there is no fundamental or harmonic frequencies that are whole-number multiples of the fundamental. It would be difficult to discern a particular pitch for a sound that is produced by turbulent modulation because of the wide spread of energy throughout the spectrum in relatively equal amounts.

Unlike vibratory modulation (with its repeated waveshape pattern over time, its signal's perceived tonal quality, and its fundamental frequency being the component with the greatest intensity that is easily identified in terms of a pitch value), the waveform for turbulent modulation indicates that a continuous sound signal is produced but that a particular waveshape is not formed nor repeated over time. In addition, the nonrepetitive pattern of the waveform leads to a signal that is perceived as **noise** (Figure 5-15B).

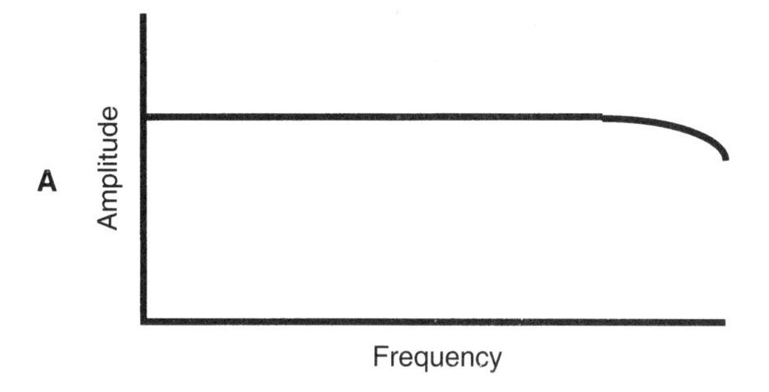

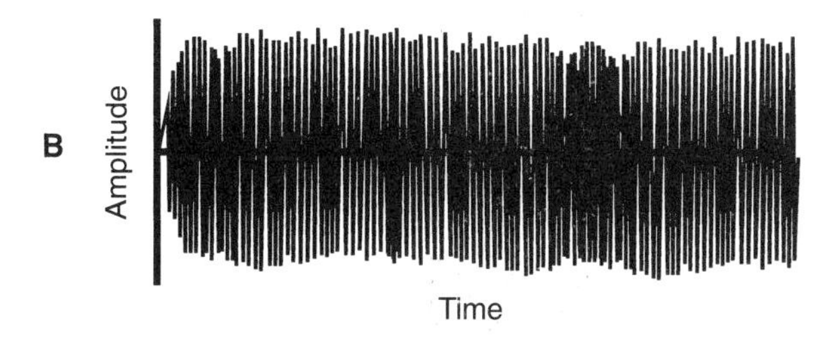

FIGURE 5-15 Spectrum (A) and Waveform (B) for Turbulent Modulation

Stepwise Modulation

Stepwise modulation, like turbulent modulation, is the result of the voiceless sound source; it is important in the production of speech sounds referred to as stops or **plosives.** These are noncontinuous noise-like sounds such as /t/ (in **t**op). Stepwise modulation is produced by the formation of a momentary complete constriction somewhere within the vocal tract (larynx to lips) (Figure 5-16). The steady airstream coming from the lungs builds up behind the constriction, which is then opened, providing a sudden expulsion of air (explosion). Stepwise modulation of the exhaled airstream results in a period of silence followed by noise comparable to that provided by turbulent modulation.

The spectrum for stepwise modulation would be similar to that for turbulent modulation if sampled during the air release phase of stepwise sound production. However, if a spectral sample were taken during the buildup phase of stepwise sound production, there would be nothing to record, because there is only silence. The release phase provides a spec-

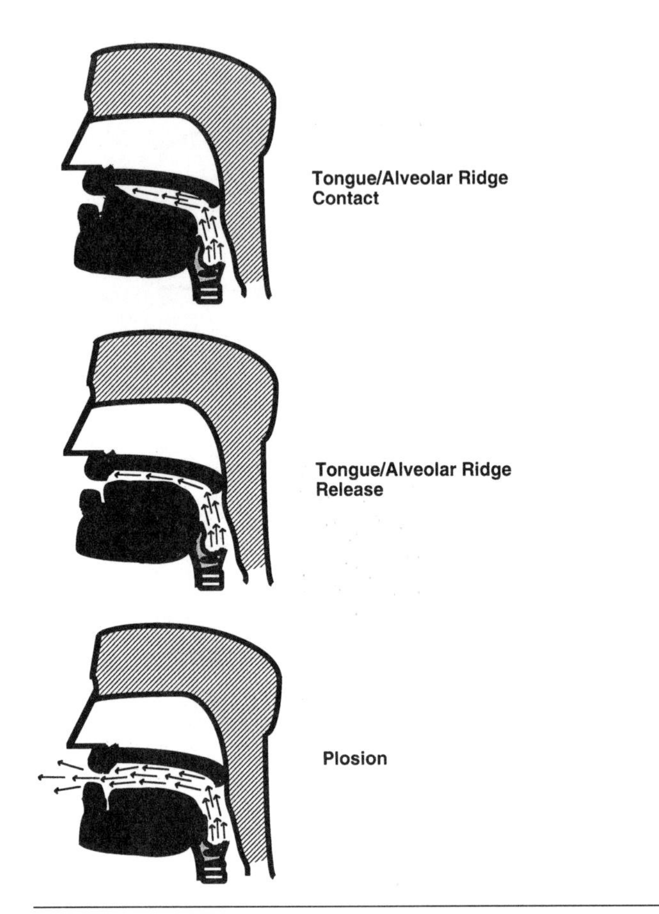

FIGURE 5-16 Stepwise Modulation Showing Constriction Created by Tongue in Vocal Tract for the Stop Plosive /t/

trum that is similar to that of white noise (Figure 5-17A). The energy spread throughout the frequency range is shown through the use of a spectral envelope, and, unlike vibratory modulation, there are no fundamental or harmonic frequencies to be seen.

The waveform for stepwise modulation indicates that a noncontinuous sound signal is produced and that during the noise phase, no particular waveshape is formed nor repeated over time (Figure 5-17B). There is a period of silence followed by a noise pattern similar to that seen for turbulent modulation. The signal that is perceived for stepwise modulation is one of silence followed by sudden noise.

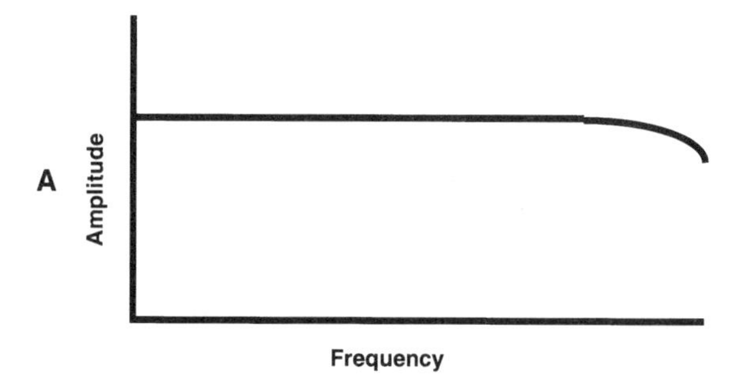

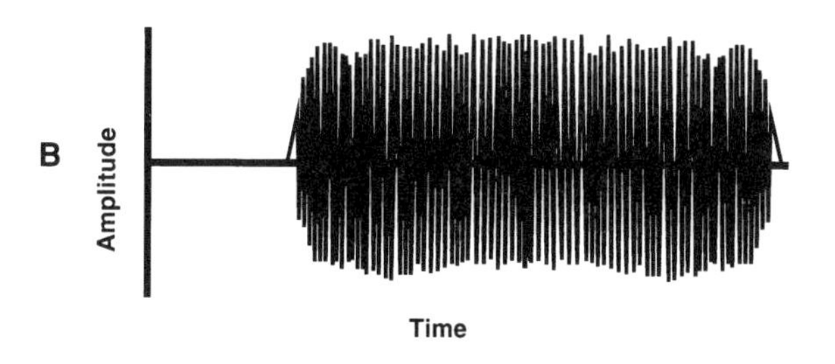

FIGURE 5-17 Spectrum (A) and Waveform (B) for Stepwise Modulation

Component #3: Resonance Source

The resonance source is the third of the three components of the speech production mechanism as it is viewed acoustically. It is analogous to a tube (vocal tract) extending from the vocal folds of the larynx to the lips. Anatomically, it includes the upper portions of the larynx, the pharynx behind the oral cavity, the oral cavity itself, and the opening between the lips (Figure 5-18). In the case of nasal sounds, the tube analogy would be modified to include two tubes, one of which would be the nasal cavity (Figure 5-19). The nasal cavity would be coupled to the oral cavity via the velum, which, for nasal sounds, would not be raised to the back wall of the pharynx (i.e., no velopharyngeal closure) (Figure 5-20).

Alteration of vocal tract shape occurs for the production of each of the speech sounds of a particular language. This shape change of the vocal tract allows it to respond differently to the excitation supplied by the sound sources discussed previously. The sound sources activate or excite the vocal tract in a vibratory, turbulent, or stepwise manner, and, in turn, the vocal tract provides certain resonance characteristics to those

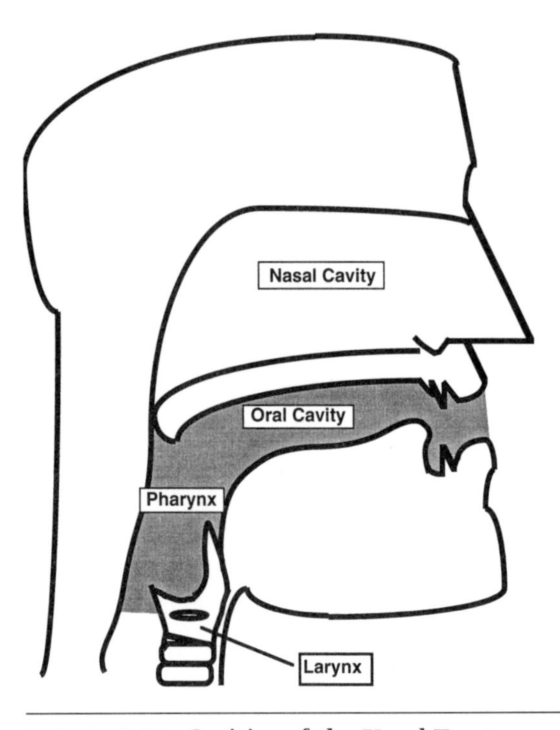

FIGURE 5-18 **Cavities of the Vocal Tract**

activations/excitations, depending on its shape at the time of activation or excitation.

The resonance source is independent of the sound sources used in speech production and can therefore be altered while the sound sources remain the same. For example, if a speaker changed the shape of his vocal

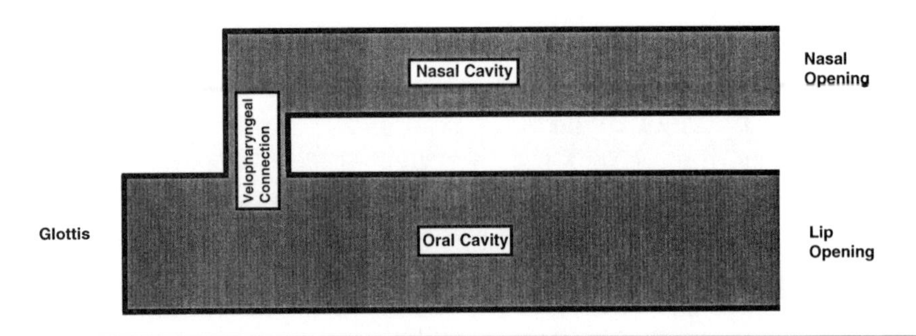

FIGURE 5-19 **Schematic of Oral–Nasal Coupling**

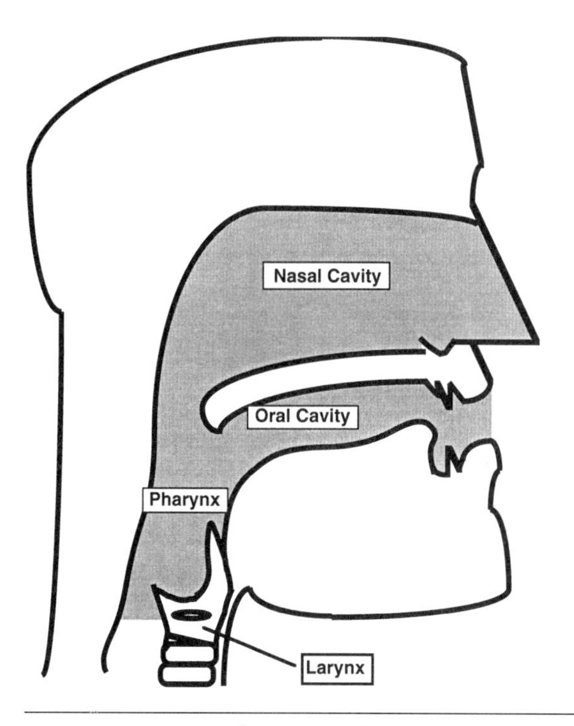

FIGURE 5-20 Vocal Tract with Nasal Coupling

tract, the speech sound being produced might shift from /i/ (as in b**ee**) to another vowel such as /u/ (as in b**oo**t), even if vocal fundamental frequency and intensity remained unaltered (Figure 5-21). The vocal pitch would be perceived as the same, but the acoustic characteristics (specifically, the formant [resonant] frequencies) and, consequently, the nature of the speech sound, would be altered; that is, the sound produced would be a different one. Similarly, if a speaker were producing a vowel such as /ə/ (as in **a**mount), he would be able to alter his vocal fundamental frequency (by modifying the rate of vibration of his vocal folds) while producing the same vowel /ə/ (by not altering the shape of his vocal tract; that is, by not altering articulatory movements in the vocal tract) (Figure 5-22). The perceived pitch of his voice would be changed, but not the formant (resonant) frequencies nor, consequently, the nature of the speech sound; listener perception of the sound itself would remain unchanged. This independence between the resonance source and the sound sources enables speakers to "articulate" the various speech sounds of a language while at the same time altering sound source patterns in terms of vocal frequency, intensity, and time, to provide additional information with

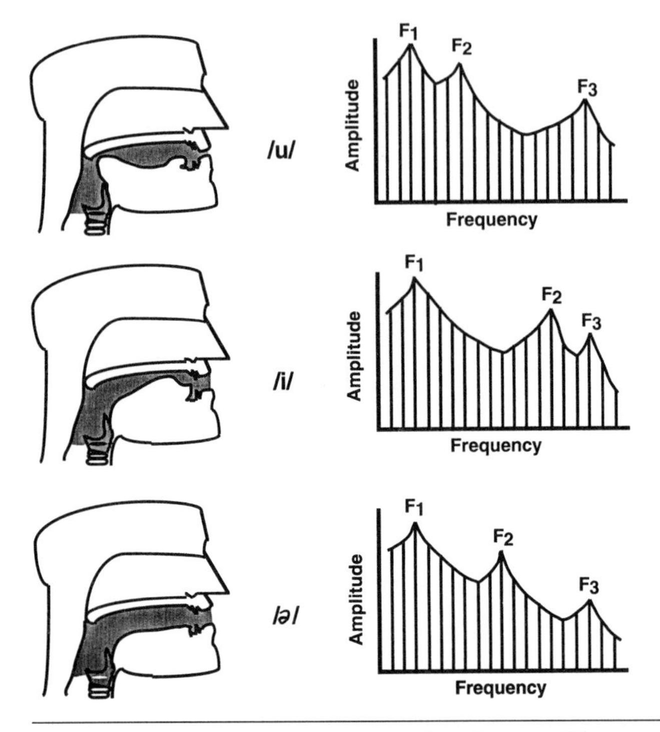

FIGURE 5-21 Vocal Tract Shapes for Three Different Vowels

regard to the communication process. Moreover, this independence of the sound sources and resonance source holds true even for voiceless sound sources that do not involve vocal fold vibration.

The simplest way to view the vocal tract as a resonant tube for speech production is by specifying its shape for the production of vowels, particularly the schwa vowel /ə/ (as in **a**bout) because of its neutral position low in the oral cavity (Figure 5-23). For the production of vowels, the vocal tract would be activated or excited by the voiced sound source (vibratory modulation). For this vowel, the vocal tract would be shaped like a tube that is closed at one end (laryngeal end), open at the other end (lip end), and multiply resonant. The tube's lowest (first) resonant frequency is equal to the frequency of a sound wave whose wavelength (λ) is four times the length of the tube ($4/1 \times L$). Its other, higher resonant frequencies are based on odd-numbered multiples (1:3:5, etc.) of its lowest resonant frequency; that is, the tube's second resonant frequency is equal to the frequency of a sound wave whose wavelength (λ) is 4/3 times the length of the tube ($4/3 \times L$). For the tube's third resonant frequency, wavelength (λ) is equal to $4/5 \times L$; the fourth resonant frequency has a wavelength (λ) of $4/7 \times L$, and so on. These rules pertain to the air particles (molecules)

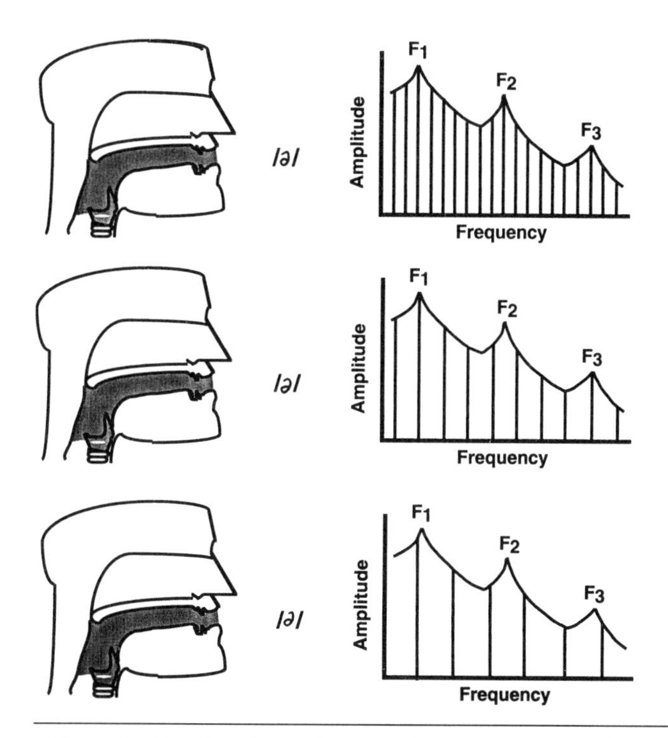

FIGURE 5-22 Vocal Fundamental Frequency Variations for the Same Vowel

located within the tube itself; the assumption is that they are being activated or excited as a result of vibratory modulation emanating from the vocal folds of the larynx. The kind of excitation of the vocal tract that is supplied by vibratory modulation is referred to as **plane wave excitation.** Plane wave excitation presents a flat wave front as it travels along the length of the vocal tract from the glottis to the lips (Figure 5-24). As was mentioned in an earlier chapter, sound emanates spherically from the

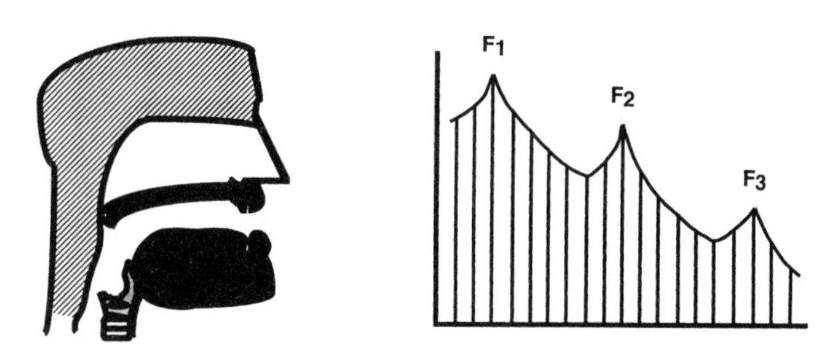

FIGURE 5-23 Vocal Tract Shape and Spectrum for the Schwa Vowel /ə/

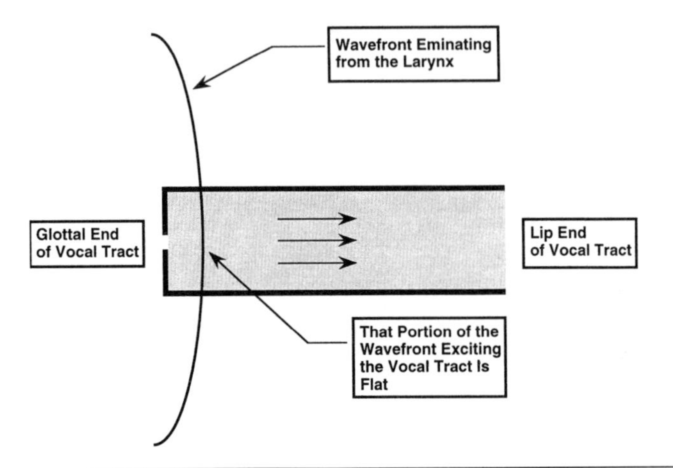

FIGURE 5-24 Plane Wave Excitation of the Vocal Tract

sound source. However, that portion of the wave front that enters the vocal tract appears to be relatively flat because of the actual diameter of the vocal tract itself. If the vocal tract were larger in diameter than it is, then the wave front exciting it might appear to be more spherical and the appropriate particle velocity and pressure patterns needed for resonance to take place would not be established. If the inanimate tube is excited by plane wave propagation, then it will have resonant frequencies that are analogous to those found in the human vocal tract during vowel production. These statements are referring to the air molecules located within the tube itself; the assumption is that they are being excited by plane wave propagation that is the result of vibratory modulation emanating from the vocal folds of the larynx.

The actual resonant frequency values of the inanimate tube analogy of the human vocal tract can be calculated. The formula for calculation of the resonant frequencies for a tube closed at one end, open at the other, and uniform in cross-sectional dimensions is

$$f_n = \frac{v}{\lambda_n}$$

$$(\lambda_n = \frac{4}{2n - 1} \times L)$$

where f is resonant frequency (equivalent to human vocal tract formant frequency), n is the particular resonant frequency that is to be calculated, v is a constant—the speed of sound in air (velocity), 4 is a constant, and L is tube length, which is critical to the resonant characteristics of a tube.

This resonant frequency formula can be used in the following examples to determine the three lowest resonant frequencies for a tube closed at one end, open at the other end, and uniform in cross-sectional dimensions. For the following examples, it will be assumed that the tube analogy involves an average adult male whose vocal tract is shaped for the production of the schwa vowel /ə/ (see Figure 5-23). It is also assumed that the tube is being activated/excited by a voiced sound source that will produce vibratory modulation within a frequency range of 100 to 3,000 Hz, which is a reasonable frequency range for vibratory modulation of the energy source produced by an average adult male speaker.

Example 1

What is the first resonant frequency of a 17-cm tube open at one end, closed at the other end, and of uniform cross-sectional dimensions throughout its length?

$$f_1 = \frac{v}{\lambda_1} = \frac{34{,}000\ cm/sec}{4 \times 17\ cm} = \frac{34{,}000\ cm/sec}{68\ cm} = 500\ Hz$$

Thus, the first resonant (formant) frequency for the average adult male is 500 Hz.

Example 2

What is the second resonant frequency of this 17-cm tube open at one end, closed at the other end, and of uniform cross-sectional dimensions throughout its length?

$$f_2 = \frac{v}{\lambda_2} = \frac{34{,}000\ cm/sec}{4/3 \times 17\ cm} = \frac{34{,}000\ cm/sec}{22.7\ cm} = 1{,}500\ Hz$$

Thus, the second resonant (formant) frequency for the average adult male is 1,500 Hz.

Example 3

What is the third resonant frequency of this 17-cm tube open at one end, closed at the other end, and of uniform cross-sectional dimensions throughout its length?

$$f_3 = \frac{v}{\lambda_3} = \frac{34{,}000\ cm/sec}{4/5 \times 17\ cm} = \frac{34{,}000\ cm/sec}{13.6\ cm} = 2{,}500\ Hz$$

Thus, the third resonant (formant) frequency for the average adult male is 2,500 Hz.

This tube analogy of the human vocal tract may also be applied to describing the vocal tract of an average adult female; the primary difference would be one of length. Because females, on average, are physically smaller in overall dimensions, it would be correctly assumed that the vocal tract for most females would be shorter than those of males. Assuming the vocal tract of the average adult female to be approximately 15 cm in length, the first resonant (formant) frequency (f_1) would be 567 Hz (34,000 cm per sec/4 X 15 cm = 34,000/60); the second (f_2) would be 1,700 Hz (34,000 cm per sec/4/3 X 15 cm = 34,000/20); and the third (f_3) would be 2,833 Hz (34,000 cm per sec/4/5 X 15 cm = 34,000/12). In every instance, the resonant (formant) frequencies of the average adult female speaker are higher than those of the average adult male (because of the length of tube factor). However, the speech sound produced, whether by a male or female, remains the same, assuming the same tube shape and a similar vibratory sound source.

Vocal tract length, therefore, has a critical effect on the resonant (formant) frequencies of the human vocal tract. The shorter the length of the vocal tract, the quantitatively higher the resonant (formant) frequencies, and, conversely, the longer the length of the vocal tract, the quantitatively lower the resonant (formant) frequencies. During production of the same vowel sound, the resonant (formant) frequencies for adult males would be lower than those for adult females, and those for adult females would be lower than those for children (because children have shorter vocal tracts than do adults). Therefore, the length of the vocal tract will affect its resonant (formant) frequencies, but it will not affect the particular speech sound being produced.

Apparently, the human ear can compensate for the different formant frequencies being produced by vocal tracts of varying lengths in order to maintain the integrity of the speech sound being perceived. There are a number of possible cues that the ear might be using, one of

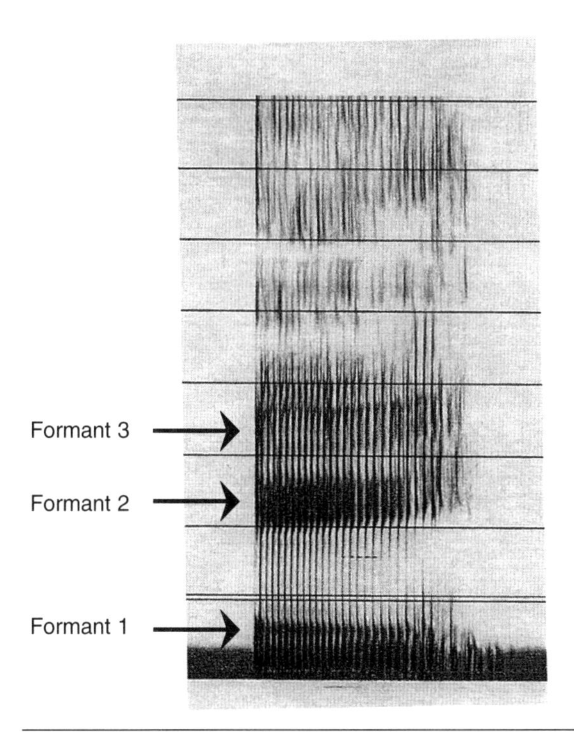

FIGURE 5-25 Spectrogram of /i/

them being the constant relationship between the resonant (formant) frequencies during the production of a particular vowel sound. For example, The vowel /i/ (as in b**ee**t) will have a low first formant frequency and high second and third formant frequencies, whether it is produced by adult males, adult females, or children. This relationship between the three lowest resonant frequencies is invariant for the production of each vowel (Figure 5-25).

So far, the descriptions of the vocal tract as a tube closed at one end and open at the other end have been restricted to a tube that is also uniform in cross-sectional dimensions (i.e., one with no constrictions) (Figure 5-26). This tube analogy applies to the production of the schwa vowel /ə/ because of its neutral position in the vocal cavity. However, in order to expand this analogy to the production of other vowel sounds, tube shape must be considered. *Shape* in this instance refers to the insertion of a constriction at some point along the length of the tube. The next logical step is to extend the tube analogy of the human vocal tract to

FIGURE 5-26 Tube Model of Human Vocal Tract

determine if its resonant behavior is predictable when it is no longer uniform in cross-sectional dimensions.

As has been suggested previously, to change the shape of a resonant tube closed at one end and open at the other end, a constriction must be inserted at some point along the length of the tube (Figure 5-27). Such a constriction will have an effect on the tube's resonant frequencies, assuming the use of an appropriate sound source for excitation of the tube (vibratory modulation). Anatomically, a constriction refers to a "humping" of the tongue in the oral cavity. The tongue is capable of constricting the vocal tract either by the elevation of its tip (front portion) or body (posterior mass behind the tip) (Figure 5-28).

This tube model of the vocal tract is of interest to speech scientists because of its remarkable analogy to the human vocal tract system. The human vocal tract is comprised of cavities that extend from the larynx to the lips. More specifically, it consists of the pharyngeal, oral, and nasal cavities, as well as the lips. Thus, the human vocal tract includes the major articulators: tongue, teeth, lips, hard palate, soft palate, and pharynx.

The human vocal tract system is analogous to a tube closed at one end (the larynx, when the vocal folds are adducted for the production of all voiced speech sounds), open at the other end (the lips), and uniform in cross-sectional dimensions throughout its length (when producing a neutral vowel sound, like /ə/, as in **a**bout). Excitation of the *vocal tract* can be accomplished by means of *vocal fold* vibrations within the larynx. The

FIGURE 5-27 Constriction in Tube Model of Human Vocal Tract

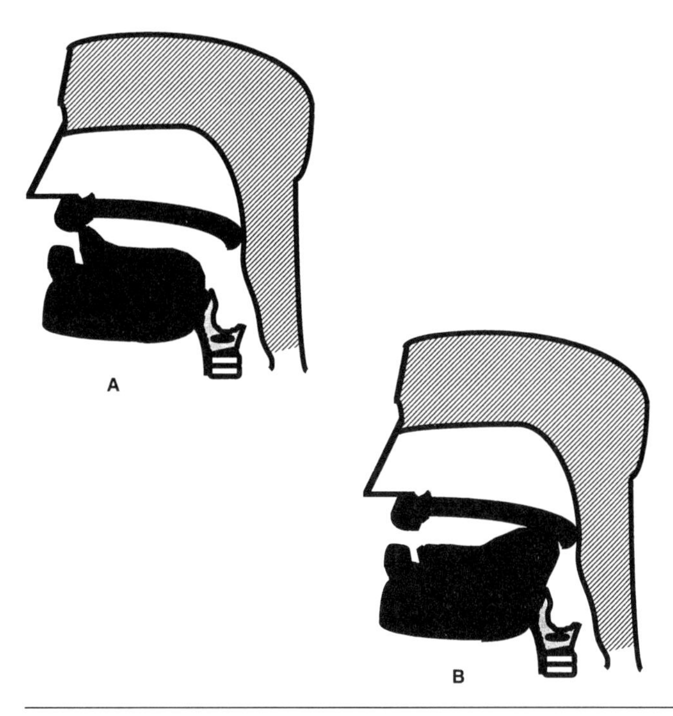

FIGURE 5-28 Maximum Constriction of Tongue in Vocal Tract for (A) a Lingua-Alveolar Consonant /t/, and (B) a Lingua-Palatal Consonant /k/

larynx provides vibrations that contain more than one frequency and stimulates the vocal tract from the tract's "closed end" (Figure 5-29).

The vocal tract of the average adult male is approximately 17 cm in length when measured from the vocal folds to the lips. If this 17-cm vocal tract is shaped so that it is of uniform cross-sectional dimensions throughout its length (i.e., minimal constriction like that for the production of the schwa vowel /ə/ [as in **a**bout]), it is analogous to a tube system closed at one end, open at the other end, and uniform in cross-sectional dimensions throughout its length. When excited by the complex, quasiperiodic, multifrequency sound source generated at the larynx, this vocal tract shape will yield the first three resonances within the tube at approximately 500 Hz, 1,500 Hz, and 2,500 Hz. *Note:* These are the very same frequencies derived from the acoustic theory's tube model of the vocal tract based on the lawful behavior of a column of air inside an inanimate cavity or tube of 17 cm in length.

The length of the vocal tract is an important factor in the phenomenon of resonance. As was the case with the tube system described pre-

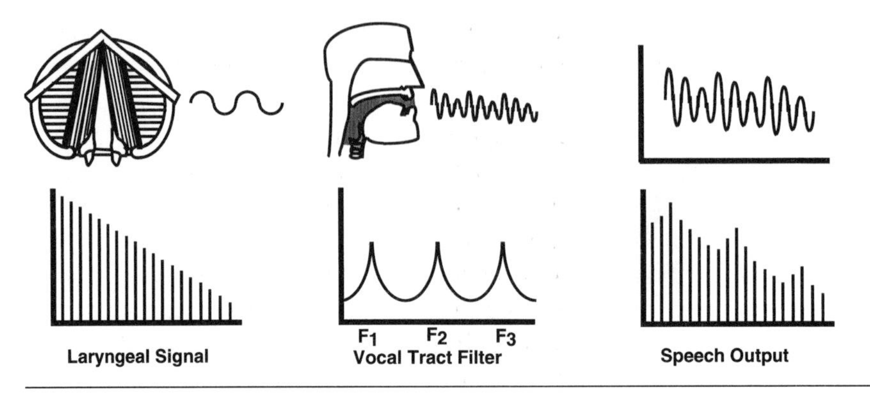

FIGURE 5-29 Effects of Acoustic Filtering

viously, alteration of vocal tract length will affect its natural, resonant frequencies. On average, adult males have longer vocal tracts than do adult females, and adult females, in turn, have longer vocal tracts than do children. Therefore, it would be expected that production of a given vowel by adult female speakers would be accomplished with higher resonant frequencies than those for adult males, and, similarly, the production of the same vowel by children would involve higher resonant frequencies than those for women. In each case, the same vowel sound would be heard, but the resonant (formant) frequencies of the respective vocal tracts would be different, depending on their overall length (Table 5-1).

TABLE 5-1 ***Average Formant Frequences (Hz) of Vowels by 76 Speakers***

		i	ɪ	ɛ	æ	ɑ	ɔ	ʊ	u	ʌ	ɝ
F_1	Adult male	270	390	530	660	730	570	440	300	640	490
	Adult female	310	430	610	860	850	590	470	370	760	500
	Child	370	530	690	1010	1030	680	560	430	850	560
F_2	Adult male	2290	1990	1840	1720	1090	840	1020	870	1190	1350
	Adult female	2790	2480	2330	2050	1220	920	1160	950	1400	1640
	Child	3200	2730	2610	2320	1370	1060	1410	1170	1590	1820
F_3	Adult male	3010	2550	2480	2410	2440	2410	2240	2240	2390	1690
	Adult female	3310	3070	2990	2850	2810	2710	2680	2670	2780	1960
	Child	3730	3600	3570	3320	3170	3180	3310	3260	3360	2160

(Modifed with permission from G.E. Peterson & H.L. Barney, Control methods used in a study of vowels. *Journal of the Acoustical Society of America 24,* 1952, 183.)

Therefore, for every frequency there is a certain length of a resonator that will yield maximum resonance; the lower the frequency, the longer the tube or cavity.

Once the shape of the vocal tract is altered to produce any of the speech sounds of English, the relationship between the resonances noted for the straight tube analogy no longer holds true. Thus, the human vocal tract is analogous to a tube closed at one end, open at the other end, and uniform in cross-sectional dimensions, but only when it is approximating a configuration required for the production of the neutral schwa vowel /ə/. However, once the vocal tract shape is altered, as it would be for the production of other vowel sounds (Figure 5-30), its resonant properties (formant frequencies) will also change.

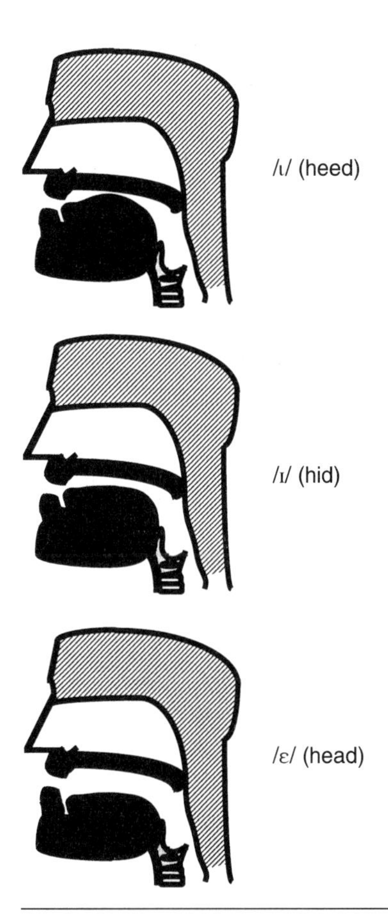

FIGURE 5-30 Maximum Constriction of Tongue in Vocal Tract for Three Different Vowels

Summary

The source–filter theory of speech production involves viewing speech production acoustically as a three-part system that is modulatory in design. The controlled exhaled air supply coming from the lungs (energy source) is altered by the voiced and/or voiceless sound sources in such a way that it becomes audible sound, which is further altered by the resonance source, so that it is converted into speech sounds.

In the case of vibratory modulation for voiced speech sounds, the air is pulsed by the rapid repetitive movement of the vocal folds housed in the larynx. For turbulent modulation, which produces the fricative consonants, the air is forced through a narrow constriction somewhere in the vocal tract, causing a wide band of noise to be emitted in a continuous manner. Stepwise modulation for the production of plosive consonants is similar to turbulent modulation, with the exception that the air supply is stopped behind a complete constriction somewhere in the vocal tract (silence) and then released, yielding a wide band of noise similar to that for turbulent modulation.

The resonance source acts like a filter in regard to the sound passing through it. Because of its shape at a particular point in time, it will emphasize some frequencies coming from the sound sources more than others. Thus, it will resonate in conjunction with those frequencies from the sound sources that are similar to its own natural frequencies (i.e., the natural frequencies at which it would vibrate if set in motion and then left to itself). Consequently, a particular speech sound that we perceive will be "coded" acoustically by the frequency arrangements unique to it.

The concept of formants and their relationship to the shape of the vocal tract was discussed in detail, including constrictions in the vocal tract created by movements of the articulators. The shape of the vocal tract for vowel production depends primarily on the insertion of a constriction somewhere along its length. Although the formants (resonances) will differ in frequency when the same vowel is produced by adult males, adult females, and children, the relationships between the formants will remain relatively the same. These acoustic cues, plus the listener's knowledge of the language, will enable a speaker to be understood, even though each speaker might have a vocal tract of different length. Because of the listener's knowledge of the language being spoken, it is even possible for the speaker to be understood when she is providing a speaking rate that is too fast for the target positions (precise tongue placement) for individual vowels to be established in continuous speech. The actual formant frequencies that are usually required for a particular vowel to be produced might not be established in rapid speech, but the listener can fill in the missing details through contextual information and the knowledge of what the formants should be for a particular vowel production.

Study Questions

1. Using the acoustical model of speech production, explain
   a. energy source
   b. sound sources
   c. resonance source
2. What is *vibratory modulation,* and what is its role in speech production?
3. Define the following acoustic concepts as well as their physiological and perceptual correlates:
   a. vocal fundamental frequency
   b. vocal intensity
4. Describe the basic premise in the *myoelastic–aerodynamic theory of phonation.*
5. What is the *Bernoulli effect* (or *Bernoulli principle*), and what is its role in phonation?
6. What is *turbulent modulation,* and what is its role in speech production?
7. What is *stepwise modulation,* and what is its role in speech production?
8. What is *coupling* and what is its role in the production of nasal speech sounds?
9. What are *formants* and why are they important in speech production and speech perception?
10. Calculate the *third formant frequency* for the average adult male vocal tract.
11. Calculate the *fifth formant frequency* for the average adult female vocal tract.
12. Using the tube model of the vocal tract, explain why differences exist in the *formant frequencies* of the same vowels spoken by *adult females, adult males,* and *children.*

Suggested Readings

Borden, G.J., Harris, K.S., & Raphael, L.J. (1994). *Speech Science Primer: Physiology, Acoustics, and Perception of Speech.* (Third Edition). Baltimore: Williams & Wilkins.

Fant, G. (1960). *Acoustic Theory of Speech Production.* The Hague: Mouton.

Fujimura, O., & Erickson, D. (1997). Acoustic phonetics. In Hardcastle, W., & Laver, J. (Eds.), *The Handbook of Phonetic Sciences.* Oxford, England: Blackwell, 65–115.

Hillenbrand, J., Getty, L.A., Clark, M.J., & Wheeler, K. (1995). Acoustic characteristics of American English vowels. *Journal of the Acoustical Society of America* 97, 3099–3111.

Kent, R.D. (1997). *The Speech Sciences.* San Diego: Singular Publishing Group.

Kent, R.D., Dembowski, J., & Lass, N.J. (1996). The acoustic characteristics of American English. In Lass, N.J. (Ed.), *Principles of Experimental Phonetics.* St. Louis: Mosby, 185–225.

Kent, R.D., & Read C. (1992). *The Acoustic Analysis of Speech.* San Diego: Singular Publishing Group.

Olive, J.P., Greenwood, A., & Coleman, J. (1993). *Acoustics of American English Speech.* New York: Springer Verlag.

Stevens, K.N., & House, A.S. (1961). An acoustical theory of vowel production and some of its implications. *Journal of Speech and Hearing Research* 4, 303–320.

6

Speech Perception

How do we perceive speech? This apparently simple question has no really simple answer. Just like the speech production process, the perception of speech is a very complicated, multifaceted process that is not yet fully understood. But there are a number of models and theories that attempt to explain this process.

Our knowledge of speech perception has been advanced by means of research that has utilized two laboratory instruments: the sound **spectrograph** and the **Pattern Playback.** The sound spectrograph, which consists of a series of analyzing filters, allows us to analyze the speech signal via wideband or narrowband "windows," and thus to determine the acoustic cues for different speech sounds. It provides information on the fundamental frequency and harmonics of the voice as well as the resonances (formants) of the vocal tract. The digital sound spectrograph yields a printout of a spectrogram that provides this acoustic information on the speech signal (Figure 6-1).

The Pattern Playback is a speech synthesizer that converts painted visual patterns into speechlike sounds. It performs the reverse of the sound spectrograph in that it converts a visual input into a perceived auditory signal. Synthetic replicas of speech are produced by hand-painting patterns representing the formants of a sound on acetate film loops, and then converting them into acoustic signals by a photoelectric system (Figure 6-2), thus assisting in the determination of the important acoustic features in the speech perception process. It is by means of synthesizing and editing the speech signal that speech scientists have been able to systematically alter various parameters of the acoustic signal and thereby determine their effect on listeners' perception. What has been learned about the perception

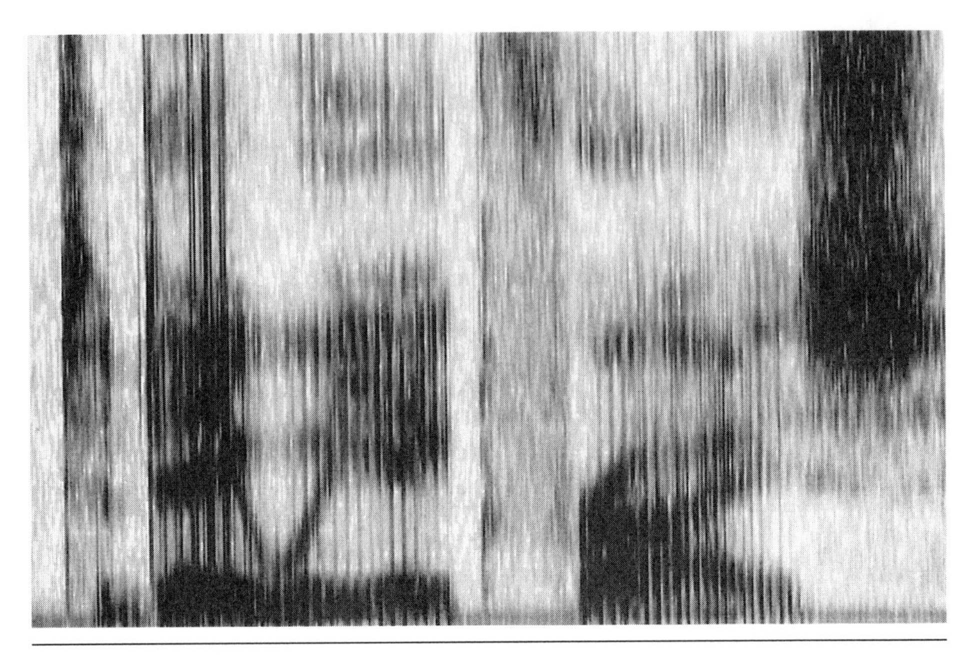

FIGURE 6-1 A Spectrogram

of speech from analysis, synthesis, and listeners' perceptual judgments of speech signals is summarized next.

Vowel Perception

Formants, resonances of the human vocal tract, are important acoustic cues for the identification of vowels. It has been found that, in general, the first two or three formants (F_1, F_2, and F_3) are sufficient for the perceptual identification and differentiation of vowels (Figure 6-3).

Diphthong Perception

Diphthongs, combinations of two vowels, exhibit **formant transitions,** which are frequency changes in a portion of the formants, reflecting changes in the shape of the vocal tract via articulator movements. Thus, formant transitions are cues for the identification of diphthongs (Figure 6-4).

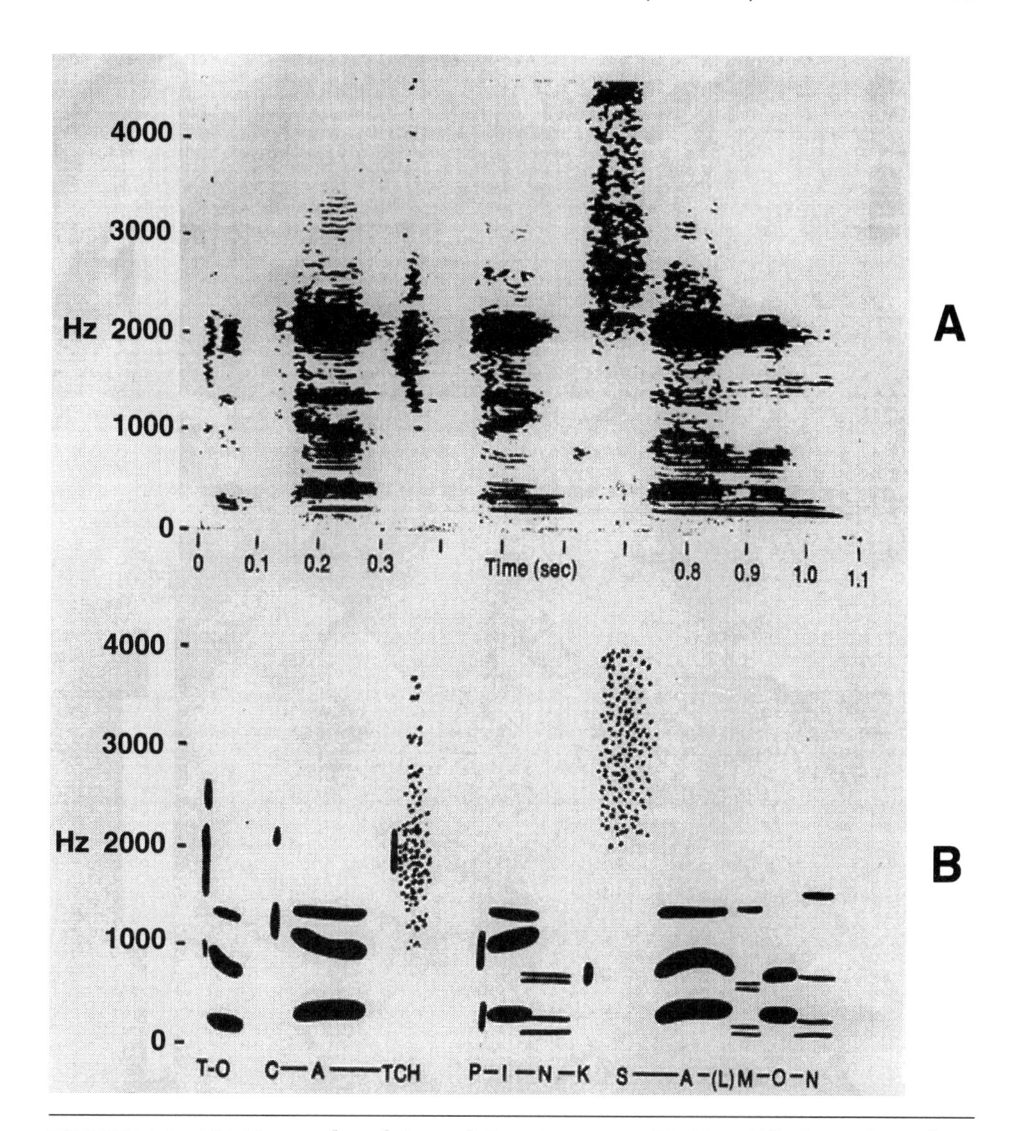

FIGURE 6-2 (A) Narrowband Sound Spectrogram. (B) Simplified Version of the Same Phrase Used in A that Serves as Input to Pattern Playback. *(Abstracted with permission from Cooper, F.S., Delattre, P.C., Liberman, A.M., Borst, J.M., & Gerstman, L.J. [1952]. Some experiments on the perception of synthetic speech sounds.* Journal of the Acoustical Society of America *24, 597–606. Copyright 1952 Acoustical Society of America.)*

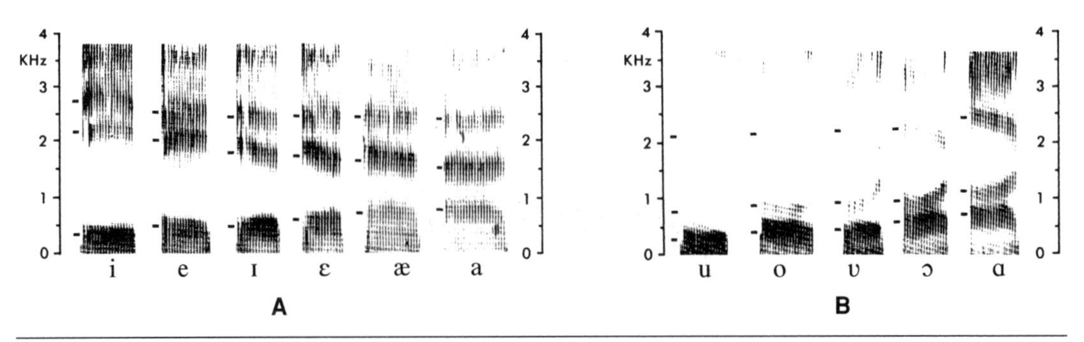

FIGURE 6-3 Spectrograms of Front (A) and Back (B) Vowels. *(From Pickett, J.M.* The Acoustics of Speech Communication: Fundamentals, Speech Perception Theory, and Technology. *Copyright © 1999 by Allyn & Bacon. Reprinted by permission.)*

Consonant Perception

The perception of consonants is more complex than vowel perception partly because consonants depend on vowels for their recognition. For example, if the recorded segments of stop consonants are separated from vowels, they will not be perceived as stops. Stop consonant perception is dependent on rapidly changing formant transitions from the consonant to the vowel in a consonant–vowel (CV) context.

While the formant transition is that segment of the acoustic signal that allows for the perception of a consonant (Figure 6-5), if the auditory

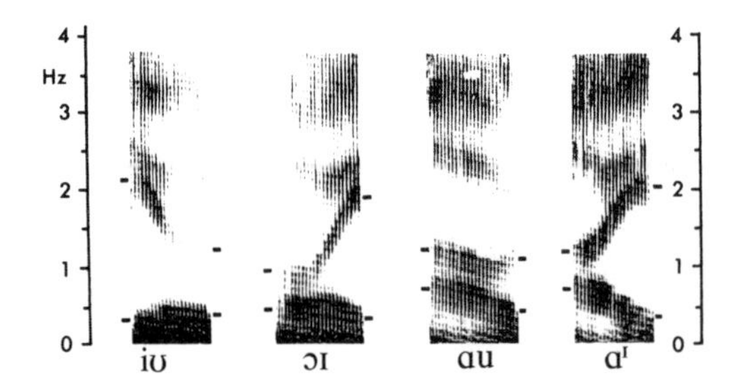

FIGURE 6-4 Spectrograms of Diphthongs. *(From Pickett, J.M.* The Acoustics of Speech Communication: Fundamentals, Speech Perception Theory, and Technology. *Copyright © 1999 by Allyn & Bacon. Reprinted by permission.)*

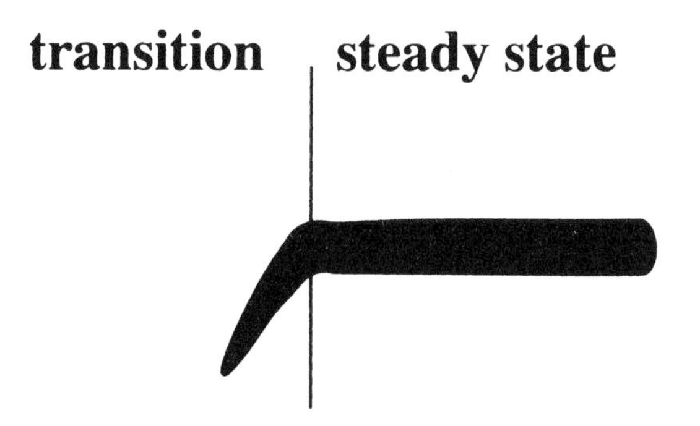

FIGURE 6-5 Formant Transition and Steady State Portion of a Consonant. *(Adapted from Ryalls, J. [1996].* A Basic Introduction to Speech Perception. *San Diego: Singular.)*

signal corresponding solely to the formant transition is presented to listeners, they will not hear an isolated consonant. Furthermore, if the portion of the tape containing the vowel sound is intentionally deleted, the consonant sound also disappears, and the remaining sound does not resemble a speech sound at all. Although the acoustic correlates of the consonant are still available in this formant transition portion of the signal, the perceptible consonant sound cannot be isolated.

While consonant perception depends on the formant transition segment of the spectrogram, this segment changes significantly when the same consonant is spoken with different accompanying vowels (Figure 6-6).

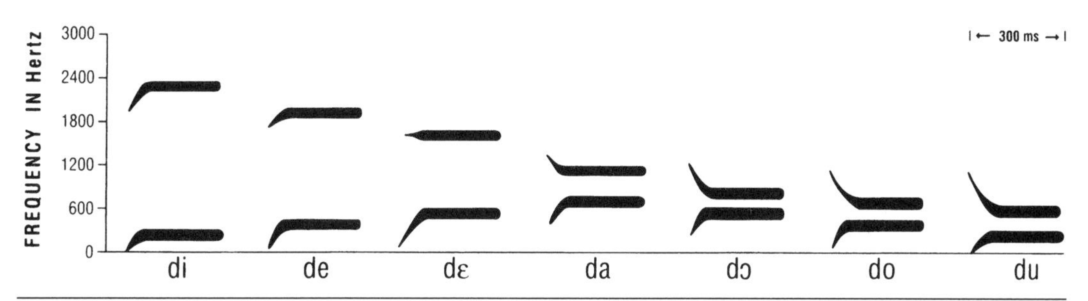

FIGURE 6-6 Schematic Representations of First Two Formant Frequency Patterns for /d/ in Front of Different Vowels. *(Adapted from Liberman, A.M., Cooper, F.S., Shankweiler, D.S., & Studdert-Kennedy, M. [1967]. Perception of the speech code.* Psychological Review *74, 431–461. Copyright © 1967 by the American Psychological Association. Adapted with permission.)*

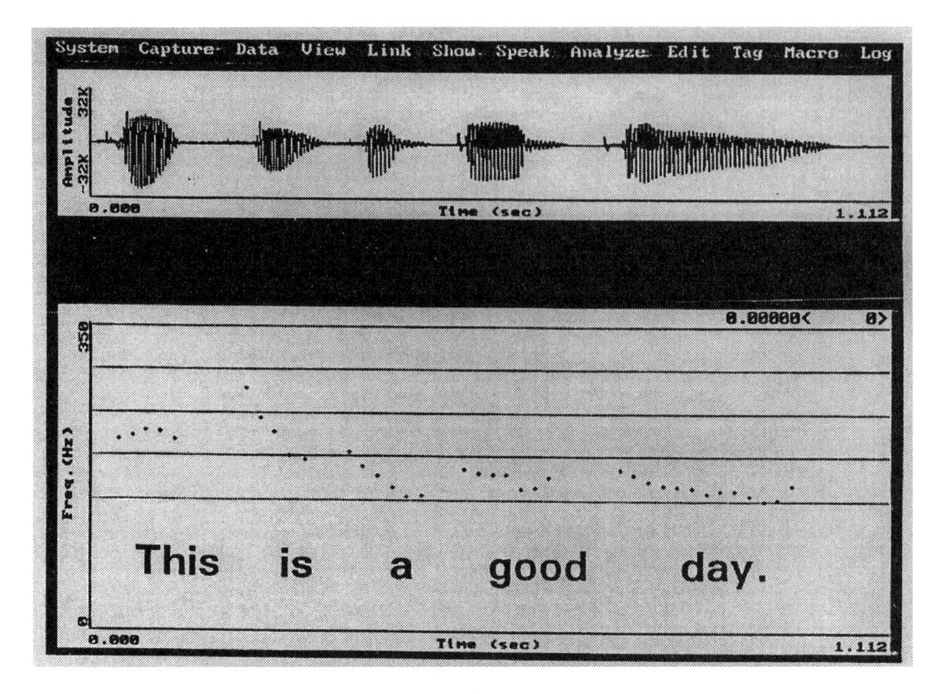

A

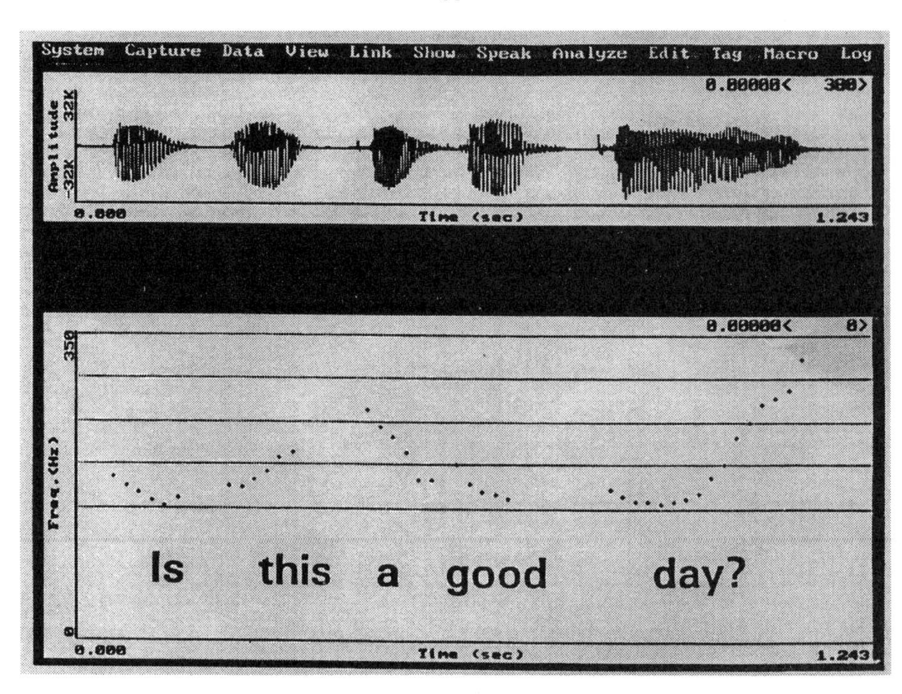

B

FIGURE 6-7 Intonation Patterns of English. (A) Declarative Statement. (B) Interrogative Statement. *(Reprinted from Kent, R.D., Dembowski, J., & Lass, N.J. [1996]. The Acoustic Characteristics of American English. In Lass, N., [Ed.],* Principles of Experimental Phonetics. *St. Louis, MO: Mosby, 215–216.*

Suprasegmental Perception

The **suprasegmental (prosodic) features** of a language are those properties of speech sounds that appear simultaneously with the phonetic (segmental) features, but are not confined to phonetic segments and, instead, are overlaid or superimposed on syllables, words, phrases, and sentences. These suprasegmental features, which include **intonation, stress,** and **quantity timing,** may alter the meaning of an utterance by revealing speakers' feelings and attitudes in a manner that phonetic features alone cannot achieve. Listener perception of suprasegmental features is dependent on the acoustic cues of vocal fundamental frequency, intensity, and/or duration or overall temporal pattern.

Intonation, which involves changes in vocal fundamental frequency, is perceived as the pitch pattern of a phrase or sentence. It can be used to express differences in the speaker's intended meaning. Declarative statements in American English are characterized by a rise–fall intonation curve, while questions are characterized by an end-of-sentence pitch rise (Figure 6-7). Thus, it is possible to utter the same word(s) but change the meaning of the utterance from a declarative statement to a question (and vice versa) by altering the employed intonation pattern.

The perception of stress, the degree of force of an utterance, involves the three acoustic parameters of intensity, duration, and fundamental frequency. Perceived stressed syllables involve increases in these three parameters. In disyllabic words, stress on the first syllable indicates a noun (for example, OBject), while changing the stress to the second syllable denotes a verb (for example, obJECT) (Table 6-1).

Quantity timing pertains to duration within a phonological system. Changes in the relative durations of linguistic units in words can change the meaning of words, while durational differences at the sentence level (changes in the tempo of speech from a neutral rate of articulation) can convey perceptual information concerning the speaker's mood. Juncture, a prosodic feature related to duration, is concerned with the manner in which sounds are joined to or separated from each other. Variations in juncture cause changes in perceived lexical meanings in English. For example, the difference between "I'm meeting" and "I'm eating," or between "In the beginning" and "In the big inning," are junctural in nature. In the first example, if the [m] is more closely linked to the preceding diphthong [aI] than to the following vowel [i], then "I'm eating" is perceived by the listener. But "I'm meeting" will be perceived if the [m] is joined to the [i] and disjoined from the preceding diphthong [aI].

TABLE 6-1 ***Vowel Durations and Intensities of 12 Speakers' Productions of the Word "Object" Produced with: (A) Stress on the First Syllable and (B) Stress on the Second Syllable***

Vowel Durations (in seconds)				*Vowel Intensities (in decibels)*			
A OBject		B obJECT		A OBject		B obJECT	
V1	*V2*	*V1*	*V2*	*V1*	*V2*	*V1*	*V2*
0.12	0.09	0.05	0.16	15	9	5	12
0.14	0.13	0.07	0.13	20	12	6	11
0.16	0.15	0.07	0.16	16	8	7	12
0.17	0.10	0.05	0.19	16	11	9	13
0.13	0.11	0.04	0.20	15	3	3	9
0.13	0.09	0.06	0.14	18	14	6	12
0.14	0.15	0.09	0.18	18	15	11	16
0.15	0.08	0.05	0.19	15	6	3	15
0.19	0.12	0.02	0.19	18	11	4	15
0.18	0.15	0.06	0.18	14	4	12	15
0.17	0.14	0.06	0.18	12	6	15	18
0.16	0.15	0.07	0.16	13	7	12	12

(From Fry, D.B. [1955]. Duration and intensity as physical correlates of linguistic stress. Journal of the Acoustical Society of America *23, 765–769.)*

Issues in Speech Perception

Invariance, Linearity, Segmentation

The basic issues that are addressed in the research literature on speech perception include *acoustic–phonetic invariance, linearity,* and *segmentation.* These issues comprise the *primary recognition problem* that has been addressed extensively by investigators: how the form of a spoken word is recognized from acoustic information in the speech waveform.

The principle of acoustic–phonetic invariance states that corresponding to each phoneme (speech sound) is a distinct set of acoustic features, so that each time a given phoneme is produced, the same acoustic cues are identifiable in the speech signal, regardless of context. The linearity principle proposes that in a spoken word, a specific sound corresponds to each phoneme, with units of sound corresponding to phonemes being discrete and ordered in a particular sequence. The segmentation principle asserts that the speech signal can be divided (and recombined) into acoustically independent units that correspond to specific phonemes.

These three principles of invariance, linearity, and segmentation imply a one-to-one connection between the acoustic and phonemic properties of sounds in words that facilitates speech perception. However, evidence from extensive research on this topic indicates that natural speech does not conform to these basic invariant conditions:

1. Acoustic cues outnumber phonemes in words.

2. Acoustic properties of a given phoneme vary in different phonetic contexts.

3. At a given point in the signal, overlapping information is present about the acoustic properties of a specific phoneme as well as the phonemes that precede and follow it.

4. The articulators move continuously in conversational speech so that the shape of the vocal tract for each intended phoneme is influenced by the shapes for phonemes that precede as well as follow the target phoneme.

5. Although we perceive speech as a series of discrete phonemes and words, the temporal boundaries between phonemes are not found consistently in the spoken message.

6. The phenomenon of **coarticulation** (the simultaneous movement of two articulators) shows the lack of clear segmentation in the speech signal.

In addition to the study of the primary recognition problem, speech perception research has also addressed other important issues.

Intra- and Interspeaker Normalization

Each time we speak, we do not use identical, invariant speech parameter values. Speakers vary speech sounds in conversational speech by altering the pitch, loudness, stress, and/or rate of their speech. In addition, they vary their production of sounds as a function of a number of sociolinguistic factors, including speaker mood and attitude, social status of the listener, and affect. Sound production also changes as a function of speaker age and gender.

Despite all of this sound production variability within as well as between speakers, listeners are still able to perceive the intended spoken message. How are listeners able to ignore the irrelevant, variable aspects of the speech signal discussed earlier and to attend to its relevant features that are associated with sounds and words? In other words, how can the perceptual constancy that is necessary for speech perception be achieved in the presence of so much variability in conditions and speakers as well as in repetitions of the same sound by the same speaker?

Minimal Unit of Perceptual Analysis

The nonlinearity of speech, the lack of acoustic–phonetic invariance in speech, and the nonsegmental nature of speech have provided another issue to address: the selection of a minimal unit of perceptual analysis. Because of the large amount of information available in the speech waveform, and the limited channel capacities of the human auditory system and auditory memory, it appears that the incoming auditory information must be encoded into some representation scheme that can be processed in an efficient manner. The usual question addressed by theories of speech perception has pertained to the selection of the "best" (or most natural) coding unit. Theories have proposed different coding units, including phonetic features, phonemes, syllables, and words.

So how does the listener convert the continuously varying speech waveform into a series of discrete representations for linguistic analysis purposes? This is the fundamental question that theories of speech perception address. However, other issues that have been researched include the specialization of speech and the problem of perceptual constancy.

The Specialization of Speech Perception

Speech perception is viewed by some (but not all) scholars as a specialized process that requires specialized neural mechanisms unique to humans. They draw this conclusion based on the results of numerous investigations. In one typical study, a continuum of synthetically produced consonant–vowel (CV) syllables ranging from /b/ to /d/ to /g/ were generated on a speech synthesizer by varying the direction and extent of the second formant transitions of vowels in graded steps (see Figure 6-6).

Tape recordings of these synthetically generated CV syllables were then employed in two different listening tests: (1) an identification task in which items were presented one at a time for labeling, and (2) a discrimination task using an ABX paradigm in which subjects heard one of the stimuli (A), followed by a different stimulus (B), and then a third stimulus (X), which was the same as one of the first two stimuli (A or B). The subjects' task was to determine whether X is like A or like B; the percentage of subjects' correct matching of X with its identical member of the A–B pair was the criterion measure of accurate discrimination performance.

Results of subjects' performance on the *identification* task indicate that there is a sharp perceptual boundary between the plosives /b/ and /d/ and

/d/ and /g/. So despite the small physical changes between adjacent stimuli used to generate a synthetic continuum of CV syllables from /b/ to /d/ to /g/, subjects' perception of the syllables shifted abruptly, falling into natural categories for the phonemes /b/, /d/, and /g/. However, on the *discrimination* task, while subjects' discrimination of stimuli from different phonemic categories (/b/ vs. /d/ vs. /g/) was almost perfect, their discrimination of stimuli *within* the same phonemic category was close to simple chance guessing accuracy. Thus, unlike nonspeech stimuli, speech sounds (for example, the plosives studied in the foregoing experiment) are "discriminated" no better than they are "identified."

Usually, listeners are capable of discriminating many more different nonspeech sounds than they can actually identify. Thus, their perception is classified as continuous (or gradual). However, this difference that listeners exhibit between discrimination and identification for nonspeech sounds does not exist for speech sounds. This unique relationship between discrimination and identification of speech stimuli is called **categorical perception.** In categorical perception, the listener does not hear a sound that gradually becomes less and less like another sound. Rather, the listener categorizes speech stimuli in such a manner that the stimuli within a particular category sound alike. He hears one sound (for example, /b/), and then his perception changes very abruptly and he hears a different sound (for example, /d/). Generally, the perception of vowels is continuous (gradual, noncategorical), while plosive consonant perception involves categorical (noncontinuous) perception.

Proponents of categorical perception point to the observation of categorical perception for speech sounds as very different from the results of research in psychophysical experiments using nonspeech stimuli. Unlike the discontinuous, categorical perception of speech stimuli, the parameters of nonspeech auditory stimuli are perceived continuously. For example, there are no sudden changes in listeners' ability to detect differences in the perceived pitch of pure tones with changes in the frequency of the pure tones. While pitch perception of pure tones is not linear (i.e., any specific change in the frequency of a pure tone does not cause a corresponding and equal change in its pitch perception), it is a continuous (noncategorical) function, with subjects' discrimination functions being considerably better than their identification functions.

Thus, while listeners tend to be aware of relatively small changes in nonspeech sounds (continuous perception), categorical perception permits them to hear the phonemes of their language without being distracted by nonessential variations within a category of phonemes. It is considered a mechanism that humans have developed by which their perceptual system has become specially adapted for the perception of speech sounds, while coping with the enormous amount of information (including irrelevant differences between different productions of the same phoneme) presented at very rapid transmission rates in human speech.

It should be noted here that the concept of speech perception as a specialized process is highly controversial. Critics of this concept point to evidence from experiments showing that human subjects can perceive continuously varying nonspeech stimuli (such as tones or beeps) categorically, thus implying that general auditory mechanisms (rather than specialized neural mechanisms) can account for categorical perception. Moreover, results of studies using chinchillas, monkeys, dogs, and birds have shown that nonhuman animals are also capable of categorical perception, further weakening the argument that the perception of speech is a uniquely human phenomenon.

Perceptual Constancy (Perceptual Invariance)

Another issue in speech perception is the variability in the speech signal. In addition to phonetic contexts and duration, other common sources of variation that have been investigated include speakers' vocal fold vibrations, vocal tract characteristics (length and shape), articulatory strategies for producing speech, speaking rate, as well as speaker gender, age, regional origin, social status, emotional state, and physical condition. Because of these factors there is considerable variability from speaker to speaker in the production of the same words and phrases. Nevertheless, human listeners are capable of accurately perceiving the speech of all intelligible speakers. However, the processes responsible for the implied perceptual compensations, as well as the question of whether perceptual compensations are employed at all for this perceptual variability, are not yet fully resolved.

Other Perceptual Phenomena

Dichotic Listening

Dichotic listening involves different auditory stimuli of similar intensity and duration presented simultaneously to both ears. Because both signals cannot be heard, which one is heard helps us learn more about the processing (perception) of auditory signals. In general, results of dichotic listening investigations indicate that for most listeners, verbal stimuli presented to the right ear are heard more frequently (and accurately) than are those presented to the left ear, a phenomenon called the *right ear advantage.* The explanation offered for this phenomenon is that, although both ears are connected to each hemisphere of the brain, the *contralateral connections* (left ear to right hemisphere and right ear to left hemisphere) are stronger (more fibers or larger connections) than *ipsilateral connections* (left ear to left hemisphere and right ear to right hemisphere). Thus, listeners are more accurate at reporting verbal stimuli presented to the right ear (from the left hemisphere) than those presented to the left ear (from the right hemisphere).

In regard to the nature of verbal stimuli, results of investigations involving dichotic listening tasks indicate a very strong and very consistent right ear advantage for consonants, but a much less clear and less reliable right ear advantage for vowels. Moreover, for some listeners (those who are not expert musicians), the right hemisphere (left ear) appears to be somewhat better than the left hemisphere (right ear) in processing certain types of musical signals.

Auditory Illusions

Auditory illusions are breakdowns in listeners' perceptual accuracy that can help us learn more about the perceptual mechanisms utilized in the speech perception process. Two common auditory illusions are addressed briefly.

Verbal Transformation Effect

This illusion involves recorded repetitions of the same auditory stimulus. It has been found that continuously repeated presentation of the same stimulus (for example, a word) to a listener will result in changes in the listener's perception of that signal. While the acoustic signal remains unaltered, the listener actually hears other words, some of which appear to sound like the auditory stimulus, while others do not.

Phonemic Restoration Effect

This illusion, the failure of listeners to be aware that a portion of the speech stimulus has been removed and replaced by a nonspeech sound (for example, a noise or a cough), is part of *auditory induction,* a more general ability to perceptually restore parts of auditory signals masked or eliminated by other sounds. Thus, the listener perceptually restores the missing sound or word because of his knowledge (for example, syntax and semantics) of the language spoken.

Theories of Speech Perception

There are several common categories into which theories of speech perception can be divided:

Passive Versus Active Theories

Passive theories consider the process of speech perception to be primarily sensory and the listener as relatively passive in this process. They stress the filtering mechanisms of the listener, with speech production knowledge playing a minor role and used only in difficult situations. In *active theories,* the process of speech perception involves some aspect of speech production, with the listener viewed as having a more active part in the speech perception process. Speech sounds are sensed, analyzed for their phonetic properties by reference to how such sounds are produced, and thereby recognized.

Bottom-Up Versus Top-Down Theories

Bottom-up theories are data-driven and assume that the acoustic signal alone provides sufficient information for the listener's perceptual decision making. *Top-down theories* assume that listeners' perceptual decision making relies heavily on higher level sources of information involving cognitive or linguistic operations, with analysis of the acoustic signal alone insufficient for the listener to make perceptual judgments. Because of the complex nature of the speech perception process, no theory is exclusively bottom-up or top-down. Therefore, this dichotomy is based on the relative degree to which a theory relies more on the physical stimulus versus higher level sources of information.

Autonomous Versus Interactive Theories

Autonomous theories assume that listeners' perceptual decision making can be made in a closed, autonomous system that contains all the necessary perceptual operations for such decisions, with no need for other sources of information, such as that provided by context. *Interactive theories,* on the other hand, assume that listeners' perceptual decision-making relies on sources of other available information outside the listener's perceptual processor, or an interaction of the substages of listeners' perceptual processing.

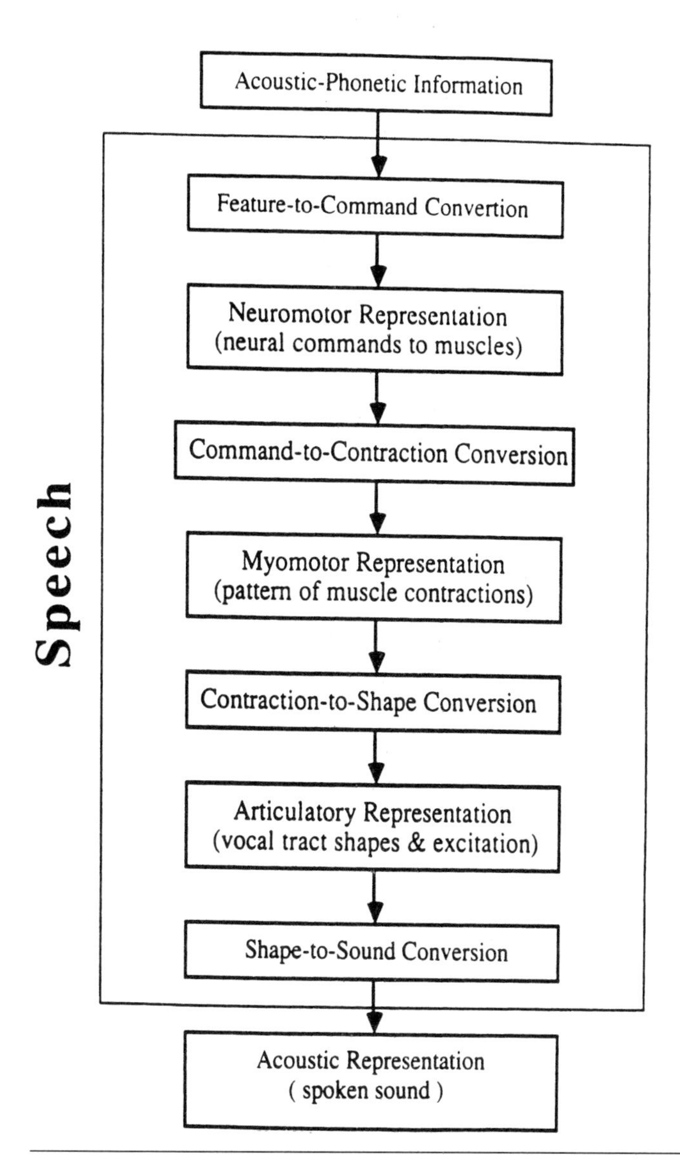

FIGURE 6-8 The Motor Theory of Speech Perception. *(Reprinted with permission from Cooper, F.S. [1972]. How is language conveyed by speech? In Kavanagh, J.F., & Mattingly, I.G. (Eds.),* Language by Ear and by Eye. *Cambridge, MA: MIT Press.)*

The following discussion of theories of speech perception is intended to provide a selected sampling of a few theories proposed to explain the complex process of speech perception.

Motor Theory

The original *motor theory of speech perception* was based on the fact that a listener is also a speaker, and thus a close link exists between the acoustic forms of speech sounds and their articulation. The underlying assumption in this theory is that speech stimuli are perceived by means of processes also involved in their production. Thus, listeners perceive speech because they are aware of how phonemes are produced by the articulators; this awareness is acquired through experience in producing phonemes and by listening to others produce phonemes (Figure 6-8).

Motor theory supports the notion of *categorical perception* for speech sounds: Speech has special perceptual properties that make it perceived differently than other, nonspeech acoustic signals, thereby suggesting the presence of a "special processor" used to decode speech. The assumption is that there are invariant motor commands to the articulators in order to produce the same phoneme in different phonetic contexts. These commands are neural messages sent by the brain to move the articulators for speech production. (This is the origin of the term *motor theory*).

In addition to the ambiguity of evidence and the highly controversial concept of speech perception as a specialized process unique to humans, evidence from investigations on infant speech perception further weakens support for the motor theory. It would seem logical that the speech perception process, like the process of speech production (as well as other motor activities), is a developmental, learned process. However, considerable evidence from studies on infant speech perception indicates otherwise: It has been found that all normal human infants are born with the innate ability to perform many of the basic processes involved in speech perception, including the ability to discriminate practically any of the sounds in all of the languages of the world. This concept runs contrary to the underlying assumption in motor theory that listeners perceive speech because they are aware of how phonemes are produced by the articulators, and this awareness is acquired through experience in producing phonemes and by listening to others produce phonemes. Newborn infants have no such experience in producing the phonemes of their future language, yet they are capable of performing many of the basic processes involved in speech perception.

Revised Motor Theory

The original motor theory of speech perception has undergone two major revisions, which are incorporated into the *revised motor theory.* These revisions are based on new data obtained since the original theory was proposed.

1. The existence of *duplex perception,* in which the same acoustic information can be processed simultaneously in both a nonspeech and speech mode. Some listeners can hear both speech and nonspeech signals simultaneously, evidence of the existence of a specialized perceptual mode in humans for speech perception purposes.

2. The notion that the brain has developed *modules,* specialized subprocessing routines in specialized areas in the brain, for addressing some perceptual information. These areas operate relatively independently of other brain processing. The perception of speech depends on these specialized innate modules.

In the revised motor theory, speech input provides the listener with the phonetic intent of the speaker, represented as an invariant gesture in the speaker's brain. The listener perceives the speaker's representation by means of an innate perceiving module that employs the acoustic pattern to retrieve the intended gesture, thereby resulting in the perception of speech. Each gesture has properties that are different from other gestures and each gesture is invariant. Therefore, a particular vocal tract configuration corresponds to a phonetically relevant category.

Analysis-by-Synthesis Theory

Like the motor theory, in the *analysis-by-synthesis theory,* speech perception is based somewhat on speech production. However, the reference in analysis-by-synthesis theory is more acoustic and less articulatory than in the motor theory and relies on a matching system. In this theory, listeners decode the acoustic signal via internally generated matching signals (a template-matching task), with the signal providing the best match as the one that the listener "perceives." According to analysis-by-synthesis theory, when the listener receives an auditory pattern, she analyzes it by eliciting an auditory model of her own production of the auditory pattern. For example, she hears an auditory speech stimulus, interprets it to be a particular phonetic message, provides a quick neural synthesis of it herself and, if there is a match of the patterns, she accepts her perception as correct.

TRACE Model

Interactive activation, a dynamic information processing system, involves excitatory and inhibitory interactions among processing units (phonemes, features, and words), as well as hypotheses about the speech input being processed.

The *TRACE model* is a connectionist model that includes attempts to account for the integration and parallel processing of multiple sources of information in speech perception. It proposes multiple levels of representation as well as feedforward and feedback connections between processing units. Processing units (called *nodes*) are arranged on three levels (phonetic feature, phoneme, and word) that, together, form a network. Each level contains detectors for the different specific dimensions of sounds. The detectors at each level are activated when input from the speaker's message is directed into the system. Perceptual processing occurs within as well as across the different levels of the system, with bidirectional connections among the levels. When information passes upward through these three levels, nodes that obtain adequate confirming evidence to exceed a threshold will fire and activate their related nodes. Figure 6-9 schematically represents the TRACE model.

An important property of the TRACE model is the organization of excitatory and inhibitory links between nodes and levels. Activation of a node on one level increases the activity of all connected nodes on adjacent levels. But within levels, all nodes are connected by inhibitory links, forcing the quick resolution of any ambiguity in the signal. For example, if the features for both /p/ and /b/ are simultaneously encountered, the nodes that correspond to the features and phonemes for both possibilities (/p/ or /b/) are activated and also inhibit their nearest competitors. Thus, the node receiving the most positive activation also receives the most "veto power" over its competitors. The TRACE model explains speech perception and word recognition in an integrated system without restrictive rules or specialized mechanisms.

Logogen Theory

In this theory, *logogens,* passive sensing neural devices associated with each word in the mental lexicon, contain all of the information about a given word, including its meaning, phonetic and orthographic (written)

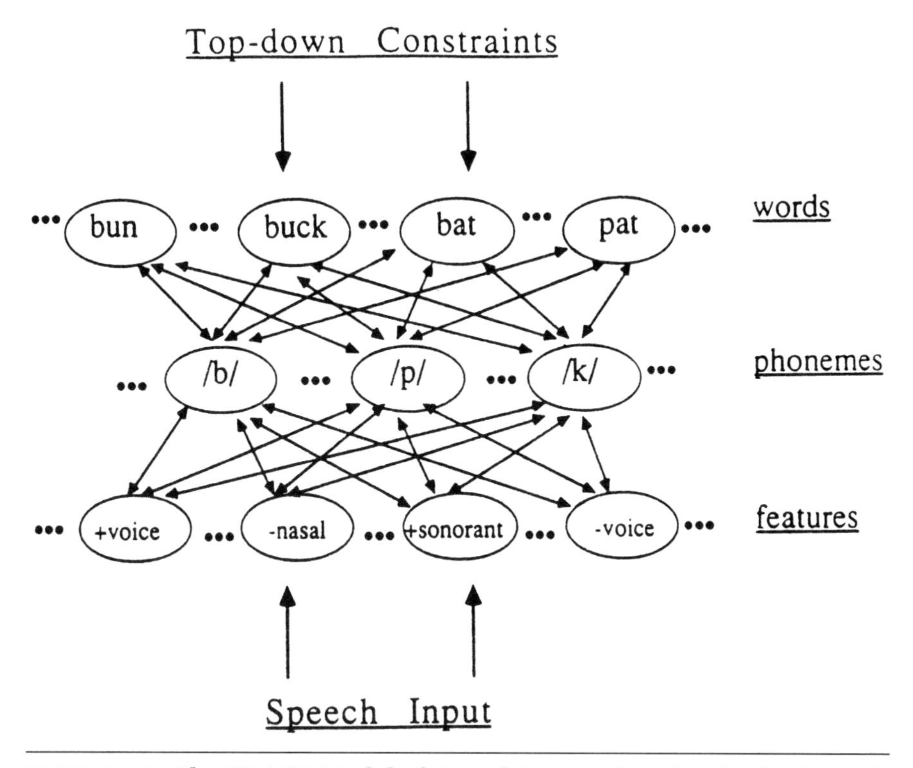

FIGURE 6-9 The TRACE Model of Speech Perception. *(Reprinted with permission from Goldinger, S.D., Pisoni, D.B., & Luce, P.A. [1996]. Speech perception and spoken word recognition: Research and Theory. In Lass, N.J. (Ed.),* Principles of Experimental Phonetics. *St. Louis: Mosby, Inc. 277–327.)*

structure, and possible syntactic functions. A logogen monitors speech production to detect any information indicating that its particular word is present in the speech signal. Once this information is detected, the logogen's activation level is raised and, if sufficiently activated, the logogen crosses a *threshold of recognition.* The information about the word that the logogen represents is made available to the response system, thereby allowing the word to be "recognized." In this theory, the emphasis is on multiple, interactive knowledge sources in word recognition. Logogens monitor all possible sources of information, and information from several levels can have a combined effect in pushing the activation level of a logogen to its threshold. An illustration of the logogen theory is shown in Figure 6-10.

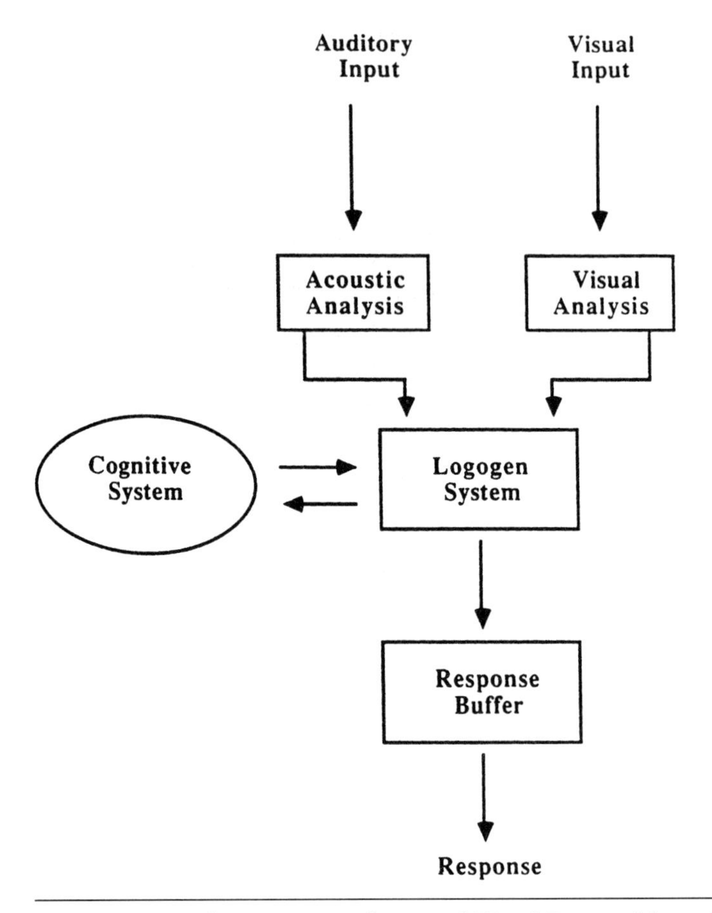

FIGURE 6-10 The Logogen Theory of Word Recognition. *(Adapted from Morton, J. [1979]. Word recognition. In Morton, J., & Marshall, J.D. (Eds.),* Psycholinguistics 2: Structures and Processes. *Cambridge, MA: MIT Press, 109–156.)*

Cohort Theory

Cohort theory proposes two stages in spoken word recognition: an *autonomous stage* and an *interactive stage*. In the first (autonomous) stage, acoustic–phonetic information at the beginning of an input word activates all words in memory having the same word–initial information. The words that are activated based on word–initial information comprise a *cohort*. For example, if the speaker uses a word that begins with the syllable "par," then the listener would select words in the cohort that begin with "par," such as *part, park, party, partake, pardon, parfait, parsley, parson,* and so on.

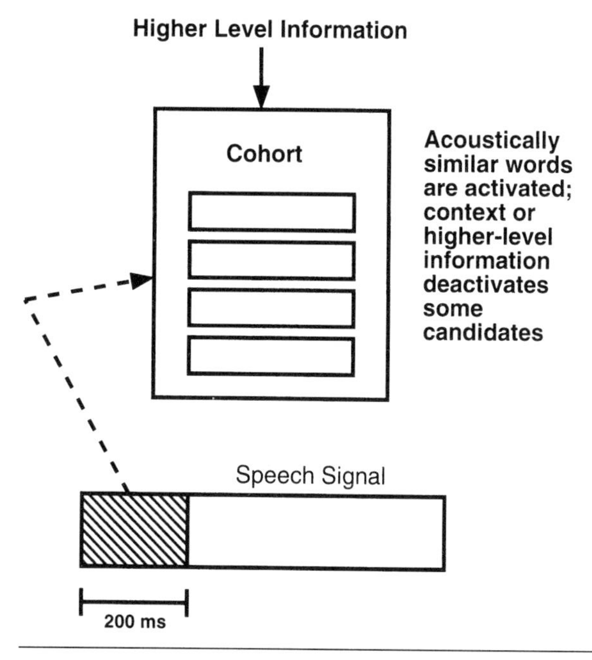

FIGURE 6-11 The Cohort Model of Word Recognition. *(Reprinted with permission by Singular Publishing Group, Inc. Kent, R. [1997].* The Speech Sciences. *San Diego: Singular, 388.)*

When a cohort is activated, inappropriate words are deactivated (eliminated as candidates from the cohort) via bottom-up information such as syntactic and/or semantic rules. All possible sources of information, including higher level knowledge sources, are involved in the selection of the appropriate word from the cohort. For example, if the word *party* does not appear appropriate in regard to the context in which it is spoken, it will be eliminated from the cohort. This is the second (interactive) stage of word recognition. Figure 6-11 illustrates the elimination process that leads to the isolation of a single word in the cohort and the achievement of word recognition.

Cohort theory is sensitive to the temporal nature of speech, giving priority to the beginnings of words and assuming a strict left-to-right processing of acoustic–phonetic information. In addition, the recognition system chooses the appropriate word candidate from the cohort at the earliest possible point, which is as soon as adequate acoustic–phonetic and higher level sources of information are consistent with a single word candidate. In cohort theory, word frequency is important in the early phases of spoken word recognition, and operates in a manner similar to the way it does in logogen theory: It modifies the individual rates of activation of the words that comprise the cohort, with more frequently occurring words becoming more active faster than less frequently occurring words.

Summary

Despite the many years and quantity of research on the perception of speech, important unresolved issues remain. In the past, the majority of research had concentrated on the perception of isolated phonetic contrasts or phonemes in short, meaningless syllables. Today the trend is toward investigating the interaction between levels of speech processing, thereby bridging the gap that traditionally has separated the investigation of the different stages of understanding spoken language. Thus, the growing interest by researchers in the perception of spoken language and the recognition of the spoken word (rather than the perception of phonetic contrasts or isolated phonemes in short, meaningless syllables) is an indication of a trend toward more comprehensive explanations of spoken language. Hopefully, future work will yield a comprehensive theory of speech perception that will resolve the currently unresolved issues and thereby provide a more complete explanation of the complex process of speech perception.

Study Questions

1. What are the important acoustic cues for the perception of
   a. vowels?
   b. diphthongs?
   c. consonants?

2. What are *suprasegmental (prosodic) features* of a language, and what role do they play in speech perception?

3. Define the following suprasegmental features in acoustic, physiological, and perceptual terms:
   a. intonation
   b. stress
   c. quantity

4. Describe the following issues and their importance in speech perception:
   a. acoustic–phonetic invariance
   b. linearity
   c. segmentation

   d. intra- and interspeaker normalization
   e. minimal unit of perceptual analysis
   f. the specialization of speech perception (versus the perception of nonspeech auditory signals)
   g. perceptual constancy (perceptual invariance)
5. What is *categorical perception,* and why is it important in speech perception? How does it differ from *continuous perception?*
6. What is *dichotic listening,* and why is it important in our understanding of the speech perception process? Include in your answer the *right ear advantage* for speech stimuli.
7. Describe the following auditory illusions:
   a. verbal transformation effect
   b. phonemic restoration effect
8. Differentiate the following categories of theories of speech perception:
   a. passive versus active theories
   b. bottom-up versus top-down theories
   c. autonomous versus interactive theories
9. Describe the following theories of speech perception (as well as compare and contrast):
   a. motor theory
   b. revised motor theory
   c. analysis-by-synthesis theory
   d. TRACE model
   e. logogen theory
   f. cohort theory

Suggested Readings

Borden, G.J., Harris, K.S., & Raphael, L.J. (1994). *Speech Science Primer: Physiology, Acoustics, and Perception of Speech.* (Third Edition). Baltimore: Williams & Wilkins.

Goldinger, S.D., Pisoni, D.B., & Luce, P.A. (1996). Speech perception and spoken word recognition: Research and theory. In Lass, N.J. (Ed.), *Principles of Experimental Phonetics.* St. Louis: Mosby, 277–327.

Greenberg, S. (1996). Auditory processing of speech. In Lass, N.J. (Ed.), *Principles of Experimental Phonetics.* St. Louis: Mosby, 362–407.

Hawkins, S. (1999). Auditory capacities and phonological development: Animal, baby, and foreign listeners. In Pickett, J.M. (Ed.), *The Acoustics of Speech Communication: Fundamentals, Speech Perception Theory, and Technology.* Boston: Allyn and Bacon, 183–197.

Hawkins, S. (1999). Looking for invariant correlates of linguistic units: Two classical theories. In Pickett, J.M. (Ed.), *The Acoustics of Speech Communication: Fundamentals, Speech Perception Theory, and Technology.* Boston: Allyn and Bacon, 198–231.

Hawkins, S. (1999). Reevaluating assumptions about speech perception: Interactive and integrative theories. In Pickett, J.M. (Ed.), *The Acoustics of Speech Communication: Fundamentals, Speech Perception Theory, and Technology.* Boston: Allyn and Bacon, 232–288.

Jusczyk, P.W. (1996). Developmental speech perception. In Lass, N.J. (Ed.), *Principles of Experimental Phonetics.* St. Louis: Mosby, 328–361.

Kent, R.D. (1997). *The Speech Sciences.* San Diego: Singular Publishing Group.

Liberman, A.M., Cooper, F.S., Shankweiler, D.S., and Stoddart-Kennedy, M. (1967). Perception of the speech code. *Psychological Review* 74, 431–461.

Ryalls, J. (1996). *A Basic Introduction to Speech Perception.* San Diego: Singular Publishing Group.

Strange, W. (1999). Perception of vowels: Dynamic constancy. In Pickett, J.M. (Ed.), *The Acoustics of Speech Communication: Fundamentals, Speech Perception Theory, and Technology.* Boston: Allyn and Bacon, 153–165.

Strange, W. (1999). Perception of consonants: From variance to invariance. In Pickett, J.M. (Ed.), *The Acoustics of Speech Communication: Fundamentals, Speech Perception Theory, and Technology.* Boston: Allyn and Bacon, 166–182.

Warren, R.M. (1996). Auditory illusions and perceptual processing of speech. In Lass, N.J. (Ed.), *Principles of Experimental Phonetics.* St. Louis: Mosby, 435–466.

Glossary

Absorption: The taking in of sound wave energy; the opposite of reflection.

Acoustic Nerve (Cranial Nerve VIII): One of 12 cranial nerves. It is responsible for connecting the hair cells of the cochlea with the auditory centers in the temporal cortex of the brain. More specifically, this nerve connects the hair cells of the cochlea with Heschl's gyrus which is located in the upper temporal lobe cortex just under the lateral fissure.

Alveolar (Gum) Ridge: A narrow bony shelf lying directly behind the upper central incisors. A major contact point for "tongue-tip" sounds.

Amplitude: The magnitude or volume of an auditory stimulus.

Aperiodicity: When the wave shape of a sound does not repeat itself over time. The sound heard is usually noise-like in nature.

Articulation: The shaping of the human vocal tract to produce different speech sounds.

Arytenoid Cartilages: Two small pyramid-shaped cartilages that serve as posterior attachments for the vocal folds.

Auditory Cortex: The portion of the brain that is responsible for interpreting auditory messages.

Autonomic Nervous System: Chains of nerve cells lying on either side of the spinal cord. The autonomic nervous system can act independently of the central nervous system when there is insufficient time for conscious decision-making to take place.

Bandwidth: The range of frequencies involved in a complex auditory signal; the range of frequencies to which a resonator responds.

Bernoulli's Principle: An elementary principle of physics which states that, in tube systems, where the speed of air flow (velocity) is greatest, the pressure on the sides of the tube wall is least.

Brain: That part of the central nervous system responsible for all behavior. The brain controls sensory input, motor output, memory, emotions, animalistic behavior, and cognitive behaviors that make us uniquely human.

Broca's Area: An area in the frontal lobe of the brain concerned with motor speech. It is responsible for movement of the intrinsic laryngeal muscles, the tongue muscles, the lip and facial muscles, and so on.

Categorical Perception: A type of perception in which a listener does not hear a sound that gradually becomes less and less like another sound, but rather the listener categorizes auditory stimuli in such a manner that the stimuli within a particular category sound alike.

Cavity (Acoustical) Resonance: The phenomenon of resonance that occurs in tubes or cavities. This phenomenon is relevant to speech production because the vocal tract is an air-filled tube.

Central Fissure (Fissure of Rolando): A deep depression in each cerebral hemisphere, which serves as a boundary marker between the frontal and parietal lobes.

Central Nervous System: The portion of the nervous system that includes the cortex, brainstem, and spinal cord.

Cerebellum: A "lobelike" structure that serves as the motor-coordinating mechanism for voluntary motor activity, including speech production. It is also involved in regulation of muscle tone and governs the sense of balance.

Cerumen: A wax produced within the external ear canal for the purpose of keeping the canal supple and clean.

Coarticulation: The simultaneous movement of two articulators.

Cochlear Nerve: The branch of the VIIIth cranial nerve that runs from the cochlea in the inner ear through the central auditory pathway to the auditory cortex in the brain.

Complex Sounds: Sounds that have energy at more than a single frequency.

Compression: *See* Condensation and Rarefactions.

Condensations and Rarefactions: Terms relating to the compression of medium particles (condensations) and spread of medium particles (rarefactions) due to sound disturbances.

Continuants: *See* Fricatives.

Corniculate Cartilages: Small upward extensions attached to the apexes of the arytenoid cartilages.

Corpus Callosum: A band of nerve fibers connecting the right and left hemispheres of the brain.

Cortex: Outer covering of the cerebral hemispheres. It consists of millions of cells that function collectively to initiate, control, and interpret behavior.

Cranial Nerves: Twelve pair of highly specialized nerves designed primarily to connect the major motor and sensory elements of the head and neck to the central nervous system.

Cricoid Cartilage: A ring-shaped cartilage that makes up the base of the larynx. It attaches directly to the trachea.

Cycle: The movement of a vibration or particle in a vibration from rest position to maximum displacement in one direction, back through rest position to maximum displacement in the opposite direction, and then back to rest position again.

Diaphragm: A large dome–shaped muscle that separates the thoracic from the abdominal cavities. The lungs rest on top of the diaphragm. A major muscle of inhalation.

Dichotic Listening: A listening condition in which two different auditory stimuli of similar intensity and duration are presented simultaneously to both ears.

Eighth Cranial Nerve (Acoustic Nerve): The cranial nerve responsible for connecting the organ of Corti to the temporal lobe of the brain.

Elasticity: The ability of a medium to resist distortion to its original shape and to the displacement of its particles.

Epiglottis: A leaf-shaped cartilage extending upward from the thyroid cartilage to the back of the tongue.

Exponent: The number to which a base number must be raised in order to equal a desired value.

False Vocal Folds: Folds of tissue lying above the true vocal folds of the larynx. The false vocal folds do not approximate for the purpose of sound production.

Formant Transitions: Formant frequency changes that occur when the speech production mechanism changes from the production of one phoneme to another phoneme (for example, consonant to a vowel or vowel to a consonant). Formant transitions are important acoustic cues for the identification of diphthongs and semivowels.

Formants: The resonant frequencies of vowels that directly relate to the vocal tract shape for a particular vowel production.

Fourier Analysis: A mathematical system for analyzing complex periodic sounds into the individual pure tones of varying frequency, intensity, and phase of which they are composed.

Frequency: The number of complete cycles that occur during a certain time period, usually one second.

Frequency Response Curve: A graph line showing changes in frequency or frequencies over time; a graph showing the frequencies to which a resonator will respond.

Frequency Theories: A class of theories of hearing in which the analysis of the frequency and intensity of sounds is conducted in the auditory cortex, not in the cochlea. In these theories, the cochlea serves only as a transducer (converter) of mechanical energy into a neurological code (which is interpreted by the central auditory nervous system).

Fricative: A manner of consonant production signified by air being forced through a narrow constriction located within the vocal tract.

Frontal Lobe: A lobe of the brain primarily responsible for higher biological thought processing (problem solving) and voluntary motor activity.

Fundamental Frequency: The lowest component frequency of a complex sound. In phonation, the acoustic determinant of the speaker's vocal pitch.

Generative (Modulatory) System: A system in which all components are independent of each other and where each succeeding component changes the characteristics of the one before it. In the case of the acoustical model of speech production, the energy source is modulated by the sound sources which in turn are modulated by the resonance source. The end product is speech.

Glottis: Space between the edges of the true vocal folds.

Hard Palate: Comprises the major portion of the roof of the mouth. It consists of a mantle of bone separating the mouth from the nose. An important contact point for lingual–palatal sounds.

Harmonics: Frequencies in the voicing spectrum above the fundamental frequency. The harmonics are whole-number multiples of the fundamental frequency. They show a decrease in energy with an increase in frequency at a rate of 12 dB per octave.

Hearing Threshold Level (HTL): An arbitrary scale for measuring sound amplitude in relation to human hearing. The reference level for this scale is "0." Zero is set according to how well humans can hear pure tones of varying frequencies.

Heschl's Gyrus: The area of the temporal lobe connected to the cochlea by the VIIIth cranial nerve. It has a one-to-one relationship with the hair cells of the organ of Corti.

Homeostasis: A state of body equilibrium controlled by the autonomic nervous system.

Homunculus: A distorted "brain" picture of the human body based on number of cortical cells relegated to control of various body parts. More cortical tissue is responsible for control of the head, neck, arms, and hands, than for control of the remainder of the body.

Hyoid Bone: A horseshoe-shaped bone lying inside the mandible (jaw bone). It serves as a base for tongue attachment, and as a suspension system for the larynx, which hangs downward from it.

Hypothalamus: A singular structure lying on the midline of the brain just below and in front of the thalami. It is responsible for animalistic functions such as sex, eating, consummatory behavior, sleeping, and temperature regulation.

Inertia: A physical principle that states that a body in motion remains in motion, while a body at rest remains at rest. Inertia plays a role in the movement of molecules in a medium in the process of producing sound.

Infrahyoid Muscles: Muscles of the larynx that lie below the hyoid bone. Three of these muscles (sternohyoid, sternothyroid, and omohyoid) help to lower the larynx upon contraction, while one muscle (thyrohyoid) helps to elevate the larynx.

Inner Ear: The third and most medial portion of the peripheral auditory mechanism. The portion of the ear that contains the sense organs of hearing and body balance.

Insula (Island of Reil): A lobe of the brain primarily responsible for regulation of gastrointestinal functions.

Intonation: A prosodic (suprasegmental) feature involving perceived changes in vocal fundamental frequency.

Inverse Square Law: A law that states that there is an orderly relationship between a decrease in sound amplitude and the distance it is measured from the sound source. This law indicates that the intensity of a sound is inversely related to the square of the difference from the sound source.

Larynx: A cartilaginous-musculo-membranous structure located in the anterior part of the neck that contains two vocal folds that can be approximated and vibrated for the purpose of sound production. The organ responsible for producing the voicing needed for all vowels, diphthongs, and voiced consonants.

Lateral Fissure (Fissure of Sylvius): A deep depression in each cerebral hemisphere that serves as a boundary marker between the frontal and temporal lobes.

Lips: Comprise the orifice that makes up the opening connecting the oral cavity with the outside atmosphere. They are mostly comprised of the orbicularis oris muscle, which provides a sphincter-like action for production of labial sounds.

Logarithms: The mathematical components underlying the concept of the decibel. Logarithms allow the speech scientist to describe a wide range of sound intensities with a small set of numbers.

Longitudinal Waves: A wave in which particle motion occurs in the same direction as the movement of the wave. Sound waves are longitudinal waves.

Lungs: Paired structures that are crucial for vegetative breathing purposes. They allow for the passage of oxygen from the outside air to the body's structures and for the removal of carbon dioxide from the body to the outside atmosphere. They provide the airstream necessary for speech production purposes.

Mass: Any form of matter, including solid, liquid, or gas.

Medulla: Comprises the lower end of the brainstem. It contains life support centers concerned with respiration and circulation.

Middle Ear: The air-filled cavity lying behind the tympanic membrane (ear drum), which serves to equalize the air pressure on both sides of the ear drum.

Molecules: Small particles of matter. Air, as a medium for sound propagation, is considered to consist of many molecules which are spaced an equal distance apart.

Motor Cortex: That portion of the cerebral cortex that affects the motor neurons that innervate the skeletal musculature, and thereby influences motor activity of the face, neck, trunk, arms, and legs.

Myoelastic-Aerodynamic Theory of Phonation: Assumes that the intrinsic muscles of the larynx will approximate the true vocal folds so that air being exhaled from the lungs can cause the true vocal folds to vibrate in a periodic manner.

Natural Frequency: That frequency at which an elastic system will vibrate if set into vibration and left alone.

Noise: An auditory signal composed of complex aperiodic vibrations.

Occipital Lobe: A lobe of the brain primarily responsible for all aspects of the visual sense.

Organ of Corti: The sensory end organ for hearing consisting of tiny hair cells, which are connected via the VIIIth cranial nerve to the temporal lobe area of the brain.

Outer Ear: The most lateral portion of the peripheral auditory mechanism which consists of the auricle and the external auditory meatus (ear canal).

Parasympathetic Part of Autonomic Nervous System: Helps the body return to a normal state (homeostasis) once a dangerous or threatening situation has passed.

Parietal Lobe: A lobe of the brain primarily responsible for incoming sensations, bodily orientation in space, and control over reading and writing skills.

Pattern Playback: A speech synthesizer which was designed to read and convert to sound "speech patterns" painted on acetate film tapes.

Peak Amplitude: An amplitude measurement from baseline to maximum particle displacement in one direction only.

Peak-to-Peak Amplitude: An amplitude measurement from maximum particle displacement in one direction to maximum particle displacement in the opposite direction.

Period: The time it takes to complete one cycle of vibration; it is the reciprocal of frequency.

Periodicity: When the wave shape of a sound repeats itself over time. The sound heard is usually tonal in nature.

Pharynx: A cavity lying behind the nasal cavity, oral cavity, and upper larynx. It serves as an important link between numerous other cavities: the nasal cavity, the oral cavity, the auditory (Eustachian) tubes, the esophagus, and the larynx.

Phase: The portion of a cycle of vibration through which a vibrator has passed up to a given instant in time. The timing relationship between pure tones comprising a complex periodic sound.

Phonation: Periodic impedance (pulsing) of the air flow coming from the lungs, which takes place at the level of the larynx. It allows for the production of all voiced speech sounds.

Pitch: The perceptual (subjective) correlate of frequency.

Place Theories: A class of theories of hearing in which the cochlea is believed to serve as both a transducer (converter) of energy as well as an analyzer of the frequency and intensity of sounds.

Plane Wave Excitation: A wave whose front appears to be flat, even though wave propagation is spherical initially as it emanates from the sound source.

Planum Temporale: A convolution more prominent in the left cerebral hemisphere than in the right cerebral hemisphere; it appears to be an important language coordination center.

Plosive: A manner of production for consonants signified by an occlusion (a stopping of the airstream) in the vocal tract followed by a sudden release (plosion) or air.

Prosodic Features: A characteristic of speech production that overlies the speech sounds themselves. Stress, intonation, and quantity (timing) are common prosodic characteristics of speech production. Also called suprasegmental features.

Pure Tone: A tone with all of its energy located at a single frequency.

Quantity Timing: A component of prosodic (suprasegmental) feature concerned with time and timing.

Rarefaction: *See* Condensations and Rarefactions.

Reflection: The phenomenon of sound waves bouncing off surfaces.

Refraction (Deflection): The bending of sound waves from their path of propagation.

Resonance: The phenomenon whereby a body, which has a natural (resonant) frequency(ies) of vibration, can be set into vibration by another body whose frequency(ies) of vibration is (are) identical or very similar to the natural (or resonant) frequency(ies) of vibration of the first body.

Resonance Curve: *See* Frequency Response Curve.

Respiration: Air exchange that takes place within the lungs. The exhalation cycle of respiration can be used as the energy source for speech production.

Reverberation: Multiple and continuous reflections of a sound.

Ribs: Twelve pairs of bones that enclose the thorax by means of their attachments posteriorly to thoracic vertebrae and anteriorly to the sternum.

Semicircular Canals: The organ for body balance, consisting of three canals: superior, lateral, and posterior.

Simple Harmonic Motion (Sinusoidal Motion): Motion derived by calculating the movement of a particle around the circumference of a circle. In regard to sound production, a simple (pure) tone results from sinusoidal motion of some sort of vibrator. Pure tones have all of their energy at a single frequency.

Sine Curve (Sine Wave): A graphical representation of sinusoidal motion.

Sinusoidal Motion: Simple harmonic motion that is demonstrated by the swinging of a pendulum or the movement of the tynes of a tuning fork.

Soft Palate: Muscular protrusion extending backward from the hard palate. The soft palate is lowered for normal breathing and for the production of nasal sounds. It is raised against the posterior pharyngeal wall for the production of nonnasal sounds.

Sound: The disturbance of the particles of a medium.

Sound Pressure Level (SPL): A physical scale for measuring sound pressure. The reference level for this scale is .0002 dynes/cm^2 (Pa).

Sounding-Board Effect: An example of the application of the phenomenon of resonance. When a tuning fork or other vibrating object is placed on a resilient surface, its amplitude is increased.

Spectral Envelope: A horizontal line on an amplitude-by-frequency display (spectrum) used to refer to aperiodic sound. The resulting output is noise.

Spectrogram: A graph generated by the sound spectrograph that displays the acoustic characteristics of sounds.

Spectrograph: An instrument utilizing a series of analyzing filters that provides information on the acoustic characteristics of sounds.

Spectrum: A graphic representation of the frequency and relative amplitude of the components of complex sounds.

Spinal Cord: A column of nervous tissue that extends from the medulla to L2 (second lumbar vertebra) in the vertebral (spinal) canal. All nerves to the body's trunk and limbs emanate from the spinal cord.

Spinal Nerves: Thirty-one pairs of nerves that convey motor signals to the body from the central nervous system and gather sensory input from the body for central nervous system processing.

Sternum: A segmented bone in the anterior–superior region of the thoracic cavity that provides the anterior attachment of the ribs and thereby helps form the rib cage.

Stress: A prosodic (suprasegmental) feature relating to the production of a syllable. If the syllable is stressed, it is longer in duration, higher in frequency, higher in intensity, and the vowel formants reach their target positions.

Subglottic Air Pressure: Air pressure that builds up beneath the vocal folds when they are approximated in the midline.

Subglottic Portion (of Larynx)**:** The most inferior portion of the internal cavity of the larynx that lies between the true vocal folds and the first ring of the trachea.

Suprasegmental Features: *See* Prosodic Features.

Sympathetic Part of Autonomic Nervous System: The portion of the autonomic nervous system that helps to prepare the body for emergency situations.

Sympathetic Vibration: A phenomenon in which one vibrating object is set into vibration by another vibrating object which has the same or similar frequency.

Teeth: Important articulators because they serve as contact points for both the lips and the tongue.

Temporal Lobe: A lobe of the brain primarily responsible for the senses of smell and hearing.

Thalami: Two ovoid masses lying on the midline of the cerebral hemispheres. These structures are responsible for control of all objective behavior, such as voluntary motor activity, and for control of sensory input from all parts of the body.

Thoracic Cavity: A bony cage housing the lungs and heart. Consists of the sternum anteriorly, the ribs laterally, and the thoracic vertebrae posteriorly.

Thyroid Cartilage: A shield-shaped cartilage that serves as an anterior protector of the vocal folds, which are attached to its posterior surface.

Tongue: The major articulator. A large flexible muscle that makes up the floor of the mouth.

Trachea: The windpipe, the tube that extends from the larynx to the lungs.

Transverse Wave: A wave in which the particles of the medium move perpendicularly to the movement of the wave. Water waves are transverse waves.

Traveling Wave Theory: A place theory proposing that the cochlea acts as both an analyzer and transducer (converter) of sound energy. A traveling wave is presented to the cochlea via movements of the footplate of the stapes in the fenestra vestibuli (oval window) and moves up the basilar membrane until it reaches a point of maximum amplitude, which is determined by the frequency of the sound. When this point along the basilar membrane that yields maximal disturbance is stimulated by the traveling wave, a neural message is sent via the VIIIth nerve to the auditory central nervous system. The site of maximum stimulation along the basilar membrane changes with changes in the frequency of the sound: toward the base of the basilar membrane for high-frequency sounds, and moves progressively toward the apex for sounds of lower frequencies.

True Vocal Folds (or **vocal folds**): Two folds of muscle tissue in the larynx that can be approximated (adduction) for phonation or separated (abduction) for air passage. Each vocal fold stretches from the inner angle of the thyroid cartilage anteriorly to a respective arytenoid cartilage posteriorly.

Velocity: The speed of sound through a transmitting medium.

Velopharyngeal Closure: The separation of the oral and nasal cavities by means of raising the velum to the posterior pharyngeal wall. This closure is necessary to achieve normal oral–nasal balance in speech production.

Velopharyngeal Port Mechanism: The mechanism comprised of the velum and pharynx that connects the oral and nasal cavities. When the velum is raised to the posterior wall of the pharynx, it prevents the escape of air through the nasal cavity; when it is lowered, it allows the passage of air through the nasal cavity.

Velum: *See* Soft Palate.

Ventricle (or **ventricle of Morgagni**): A space in the internal cavity of the larynx separating the true and false vocal folds. Contains glandular tissue that produces mucus for lubrication of the true vocal folds.

Vertebral (Spinal) Column: Column of vertebrae (spinal bones) through which the spinal cord passes. Consists of 7 cervical, 12 thoracic, 5 lumbar, 4 sacral, and 3 or 4 coccygeal vertebrae.

Vestibular Nerve: The branch of the VIIIth cranial nerve that runs from the semicircular canals of the inner ear to the brain.

Vestibule: A space in the internal cavity of the larynx above the false vocal folds that extends upward to the aditus laryngis (entrance of the larynx).

Vibratory Modulation: Sound emanating from the larynx when the vocal folds are in the adducted position. Vibratory modulation is controlled by subglottal air pressure and vocal fold tension or stiffness.

Vocal Folds: *See* True Vocal Folds.

Vocal Ligament: A ligament on the free (medial) edge of each vocal fold.

Vocal Tract: The tube that extends from the vocal folds of the larynx to the lips. It modifies the sound coming from the voiced and voiceless sound sources for the production of speech sounds.

Waveform: An amplitude-by-time display of sound.

Wavelength: The distance between points of identical phase in two adjacent cycles of a wave. Wavelength is inversely related to frequency.

Wernicke's Area: An area of the temporal lobe connected to those parts of the brain concerned with memory and experience. It compares incoming auditory signals with past experiences for the purpose of semantic interpretation.

White Noise: Noise whose energy is equally distributed throughout the entire frequency range.

Bibliography

Abbs, J.H. (1996). Mechanisms of speech motor execution and control. In Lass, N.J. (Ed.), *Principles of Experimental Phonetics*. St. Louis: Mosby, 93–111.

Baer, T., Gore, J.C., Gracco, L.C., & Nye, P.U. (1991). Analysis of vocal tract shape and dimensions using magnetic resonance imaging: Vowels. *Journal of the Acoustical Society of America* 79, 799–828.

Baru, A.V. (1975). Discrimination of synthesized vowels /ɑ/ and /I/ with varying parameters in dog. In Fant, G., & Tatham, M.A. (Eds.). *Auditory Analysis and the Perception of Speech.* London: Academic Press.

Bell-Berti, F., & Raphael, L.J. (1995). *Producing Speech: Contemporary Issues.* NY: AIP Press.

Berg, J. van den (1958). Myoelastic-aerodynamic theory of voice production. *Journal of Speech and Hearing Research* 1, 227–244.

Bertoncini, J., Bijeljas-Babic, R., Jusczyk, P., Kennedy, L., & Mehler, J. (1988). An investigation of young infants' perceptual representations of speech sounds. *Journal of Experimental Psychology: General* 117, 21–33.

Blumstein. S. (1986). On acoustic invariance in speech. In Perkell, J., & Klatt, D.H. (Eds.), *Invariance and Variability in Speech Processes.* Hillsdale, NJ: Erlbaum.

Blumstein, S., Goodglass. H., & Tarter, V. (1975). The reliability of ear advantage in dichotic listening. *Brain and Language* 2, 226–236.

Blumstein, S.E., & Stevens, K.N. (1979). Acoustic invariance in speech production: Evidence from measurements of the spectral characteristics of stop consonants. *Journal of the Acoustical Society of America* 66, 1001–1017.

Borden, G.J., Harris, K.S., & Raphael, L.J. (1994). *Speech Science Primer: Physiology, Acoustics, and Perception of Speech.* (Third Edition). Baltimore: Williams & Wilkins.

Cooper, F.S. (1972). How is language conveyed by speech? In Kavanagh, J.F., & Mattingly, I.G. (Eds.), *Language by Ear and by Eye.* Cambridge, MA: MIT Press.

Cooper, F.S., Delattre, P.C., Liberman, A.M., Borst, J.M., & Gerstman, L.J. (1952). Some experiments on the perception of synthetic speech sounds. *Journal of the Acoustical Society of America* 24, 597–606.

Darwin, C.J. (1991). The relationship between speech perception and perception of other sounds. In Mattingly, I.G. and Studdert-Kennedy, M. (Eds.). *Modularity and the Motor Theory of Speech Perception.* Hillsdale, NJ: Erlbaum, 239–259.

Denes, P., & Pinson, E. (1993). *The Speech Chain: The Physics and Biology of Spoken Language.* (Second Edition). New York: W.H. Freeman.

Deutsch, L.J., & Richards, A.M. (1979). *Elementary Hearing Science*. Boston: Allyn & Bacon.

Durant, J.D., & Lovrinic, J.H. (1995). *Bases of Hearing Science.* (Third Edition). Baltimore: Williams & Wilkins.

Eimas, P., & Miller, J. (1992). Organization in the perception of speech by young infants. *Psychological Science* 3(6), 340–345.

Eimas, P., Siqueland, E., Jusczyk, P., & Vigorito, J. (1971). Speech perception in infants. *Science* 171, 303–306.

Elman, J.L. (1989). Connectionist approaches to acoustic/phonetic processing. In Marslen-Wilson, W.D. (Ed.), *Lexical Representation and Process.* Cambridge, MA: MIT Press, 227–260.

Elman, J.L., & McClelland, J.L. (1984). An interactive activation model of speech perception. In Lass, N.J. (Ed.), *Speech and Language: Advances in Basic Research and Practice.* (Volume 10). New York: Academic Press, 334–374.

Elman, J.L., & McClelland, J.L. (1986). Exploiting lawful variability in the speech waveform. In Perkell, J.S., & Klatt, D.H. (Eds.), *Invariance and Variability in Speech Processing.* Hillsdale, NJ: Erlbaum, 360–385.

Ettema, S.L., & Kuehn, D.P. (1994). A quantitative histologic study of the normal human adult soft palate. *Journal of Speech and Hearing Research* 37, 303–313.

Fant, G. (1960). *Acoustic Theory of Speech Production.* The Hague: Mouton.

Finkelstein, Y., Shapiro-Feinberg, M., Talmi, Y.P., Nachmani, A., DeRowe, A., & Ophir, D. (1995). Axial configuration of the velopharyngeal valve and its valving mechanism. *Cleft Palate Craniofacial Journal* 32, 299–305.

Flanagan, J.L. (1972). *Speech Analysis, Synthesis, and Perception.* New York: Springer.

Fry, D.B. (1955). Duration and intensity as physical correlates of linguistic stress. *Journal of the Acoustical Society of America* 23, 765–769.

Fucci, D., & Petrosino, L. (1984). The practical applications of neuroanatomy for the speech-language pathologist. In Lass, N.J. (Ed.), *Speech and Language: Advances in Basic Research and Practice.* (Volume 11). New York: Academic Press, 249–313.

Fujimura, O. (1990). Methods and goals of speech production research. *Language and Speech* 33, 195–258.

Fujimura, O., & Erickson, D. (1997). Acoustic phonetics. In Hardcastle, W.J. & Lauer J. (Eds.) *The Handbook of Phonetic Sciences.* Oxford, England: Blackwell, 65–115.

Gierut, J.A., & Pisoni, D.B. (1988). Speech perception. In Lass, N.J., McReynolds, L.V., Northern, J.L., & Yoder, D.E. (Eds.), *Handbook of Speech-Language Pathology and Audiology.* St. Louis: Mosby, 253–276.

Goldinger, S.D., Pisoni, D.B., & Luce, P.A. (1996). Speech perception and spoken word recognition: Research and theory. In Lass, N.J. (Ed.), *Principles of Experimental Phonetics.* St. Louis: Mosby, 277–327.

Grant, K.W., & Walden, B.E. (1996). Spectral distribution of prosodic information. *Journal of Speech and Hearing Research* 39, 228–238.

Greenberg, S. (1996). Auditory processing of speech. In Lass, N.J. (Ed.), *Principles of Experimental Phonetics.* St. Louis: Mosby, 362–407.

Harnad, S. (1987). *Categorical Perception.* Cambridge, England: Cambridge University Press.

Hawkins, S. (1995). Arguments for a nonsegmental view of speech perception. In Elenius, K., & Branderud, P. (Eds.), *Proceedings of the XIII Congress of Phonetic Sciences.* Stockholm: KTH and Stockholm University, 18–25.

Hillenbrand, J., Getty, L.A., Clark, M.J., & Wheeler, K. (1995). Acoustic characteristics of American English vowels. *Journal of the Acoustical Society of America* 97, 3099–3111.

Hixon, T.J., & Weismer, G. (1995). Perspectives on the Edinburgh studies of speech breathing. *Journal of Speech and Hearing Research* 38, 42–60.

Jusczyk, P.W. (1996). Developmental speech perception. In Lass, N.J. (Ed.), *Principles of Experimental Phonetics.* St. Louis: Mosby, 328–361.

Kahane, J.C. (1988). Histologic structure and properties of the human vocal folds. *Ear, Nose, and Throat Journal* 67, 322–330.

Kelso, J.A.S., Saltzman, E.L., & Tuller, B. (1986). The dynamical perspective on speech production: Data and theory. *Journal of Phonetics* 14, 29–59.

Kent, R.D. (1993). Vocal tract acoustics. *Journal of Voice* 7, 97–117.

Kent, R.D. (1997). *The Speech Sciences.* San Diego: Singular Publishing Group.

Kent, R.D., Adams, S.G., & Turner, G.S. (1996). Models of speech production. In Lass, N.J. (Ed.), *Principles of Experimental Phonetics.* St. Louis: Mosby, 3–45.

Kent, R.D., Dembowski, J., & Lass, N.J. (1996). The acoustic characteristics of American English. In Lass, N.J. (Ed.), *Principles of Experimental Phonetics.* St. Louis: Mosby, 185–225.

Kent, R.D., & Read, C. (1992). *The Acoustic Analysis of Speech.* San Diego: Singular Publishing Group.

Klatt, D.H. (1976). Linguistic uses of segmental duration in English: Acoustic and perceptual evidence. *Journal of the Acoustical Society of America* 59, 1208–1221.

Klatt, D.H. (1989). Review of selected models of speech perception. In Marslen-Wilson, W.D. (Ed.), *Lexical Representation and Process.* Cambridge, MA: MIT Press, 169–226.

Klatt, D., & Stefanski, R. (1974). How does a mynah bird imitate human speech? *Journal of the Acoustical Society of America* 55, 822–832.

Kluender, K.R. (1994). Speech perception as a tractable problem in cognitive science. In Gernsbacher, M.A. (Ed.), *Handbook of Psycholinguistics.* San Diego: Academic Press, 173–217.

Kuehn, D.P., & Kahane, J.C. (1990). Histologic study of the normal human adult soft palate. *Cleft Palate Journal* 27, 26–34.

Kuhl, P. (1979). The perception of speech in early infancy. In N.J. Lass (Ed.), *Speech and Language: Advances in Basic Research and Practice* (Vol. 1). New York: Academic Press.

Kuhl, P. (1981). Discrimination of speech by nonhuman animals: Basic auditory sensitivities conducive to the perception of speech-sound categories. *Journal of the Acoustical Society of America* 70, 340–349.

Kuhl, P. (1983). Perception of auditory equivalence classes for speech in early infancy. *Infant Behavior and Development* 6, 263–285.

Kuhl, P. (1987). Perception of speech and sound in early infancy. In P. Salapatek & L. Cohen (Eds.), *Handbook of Infant Perception* (Vol. 2). New York: Academic Press.

Kuhl, P., & Miller, J. (1975). Speech perception by the chinchilla: Voiced–voiceless distinction in alveolar plosive consonants. *Science* 190, 69–72.

Kuhl, P., & Miller, J. (1978). Speech perception by the chinchilla: Identification functions for synthetic VOT stimuli. *Journal of the Acoustical Society of America* 63, 905–917.

Kuhl, P.K., Williams, K.A., Lacerda, F., Stevens, K.N., & Lindblom, B. (1992). Linguistic experience alters phonetic perception in infants by 6 months of age. *Science* 225, 606–608.

Lass, N.J. (Ed.). (1996). *Principles of Experimental Phonetics.* St. Louis: Mosby.

Lass, N.J., McReynolds, L.V., Northern, J.L., & Yoder, D.E. (Eds.) (1988). *Handbook of Speech-Language Pathology and Audiology.* St. Louis: Mosby.

Lehiste, I. (1996). Suprasegmental features of speech. In Lass, N.J. (Ed.), *Principles of Experimental Phonetics.* St. Louis: Mosby, 226–244.

Liberman, A.M. (1996). *Speech, A Special Code.* Cambridge, MA: MIT Press.

Liberman. A.M., Cooper, F.S., Shankweiler, D.S., & Studdert–Kennedy, M. (1967). Perception of the speech code. *Psychological Review* 74, 431–461.

Liberman, A.M., Harris, K.S., Hoffman, H.A., & Griffith, B.C. (1957). The discrimination of speech sounds within and across phoneme boundaries. *Journal of Experimental Psychology* 54, 358–368.

Liberman, A., & Mattingly, I. (1985). The motor theory of speech perception revised. *Cognition* 21, 1–36.

Liberman, A.M., & Mattingly, I.G. (1989). A specialization for speech perception. *Science* 243, 489–494.

Lieberman, P., & Blumstein, S.E. (1988). *Speech Physiology, Speech Perception, and Acoustic Phonetics.* New York: Cambridge University Press.

Lofqvist, A. (1997). Theories and models of speech production. In Hardcastle, W.J., & Lauer, J. (Eds.), *The Handbook of Phonetic Sciences.* Oxford, England: Blackwell, 405–426.

Love, R.J., & Webb, W.G. (1986). *Neurology for the Speech-Language Pathologist.* Boston: Butterworth-Heinemann.

Massaro, D.W., & Oden, G.C. (1980). Speech perception: A framework for research and theory. In Lass, N.J. (Ed.), *Speech and Language: Advances in Basic Research and Practice.* (Volume 3). New York: Academic Press, 129–165.

Masterton, R.B. (1992). Role of the central auditory system in hearing: The new direction. *Trends in Neuroscience* 15, 280–285.

Mattingly, I.G., & Studdert-Kennedy, M. (1991). *Modularity and the Motor Theory of Speech Perception: Proceedings of a Conference to Honor Alvin M. Liberman.* Hillsdale, NJ: Erlbaum.

McClelland, J., & Elman, J. (1986). The TRACE model of speech perception. *Cognitive Psychology* 18, 1–86.

Miller, J.L., Kent, R.D., & Atal, B.S. (Eds.). (1991). *Papers in Speech Communication: Speech Perception.* Woodbury, NY: Acoustical Society of America.

Minifie, F.D. (1994). *Introduction to Communication Sciences and Disorders.* San Diego: Singular Publishing Group.

Morton, J. (1979). Word recognition. In Morton, J., & Marshall, J.D. (Eds.), *Psycholinguistics 2: Structures and Processes.* Cambridge, Massachusetts: MIT Press, 109–156.

Nolte, J. (1993). *The Human Brain: An Introduction to its Functional Anatomy.* St. Louis: Mosby.

Ohala, J.J. (1996). Speech perception is hearing sounds, not tongues. *Journal of the Acoustical Society of America* 99, 1718–1725.

Öhman, S. (1966). Coarticulation in VCV utterances: Spectrographic references. *Journal of the Acoustical Society of America* 39, 151–168.

Olive, J.P., Greenwood, A., & Coleman, J. (1993). *Acoustics of American English Speech.* New York: Springer Verlag.

Orlikoff, R.F., & Kahane, J.C. (1996). Structure and function of the larynx. In Lass, N.J. (Ed.), *Principles of Experimental Phonetics.* St. Louis: Mosby, 112–181.

Palmer, J.M. (1993). *Anatomy for Speech and Hearing.* (Fourth Edition). Baltimore: Williams & Wilkins.

Pastore, R.E., Li, X.-F., & Layer, J. (1990). Categorical perception of nonspeech chirps and bleats. *Perception & Psychophysics* 48, 151–156.

Peterson, G.E., & Barney, H.L. (1952). Control methods used in a study of vowels. *Journal of the Acoustical Society of America* 24, 175–184.

Peterson, G.E., & Shoup, J.E. (1966). The elements of an acoustic phonetic theory. *Journal of Speech and Hearing Research* 9, 68–99.

Pickett, J. M. (1999). *The Acoustics of Speech Communication: Fundamentals, Speech Perception Theory, and Technology.* Boston: Allyn & Bacon.

Pickles, J.O. (1988). *An Introduction to the Physiology of Hearing.* (Second Edition). London: Academic Press.

Pisoni, D.B. (1980). Variability of sound formant frequencies and the quantal theory of speech. *Phonetica* 37, 285–305.

Pisoni, D.B. (1992). Some comments on invariance, variability, and perceptual normalization in speech perception. In Ohala, J.J., Nearen, T.M., Deruing, B.L., Hodge, M.M., & Wiebe, G.F. (Eds.), *Proceedings ICSLP–92* (1992 International Conference on Spoken Language Processing) 1, 587–590.

Remez, R.E. (1994). A guide to research on the perception of speech. In Gernsbacher, M.A. (Ed.), *Handbook of Psycholinguistics.* San Diego: Academic Press, 145–172.

Remez, R.E., Rubin, P.E., Berns, S.M., Pardo, J.S., & Lang, J.M. (1994). On the perceptual organization of speech. *Psychological Review* 101, 129–156.

Repp, B.H. (1983). Categorical perception: Issues, methods, findings. In Lass, N.J. (Ed.), *Speech and Language: Advances in Basic Research and Practice.* (Volume 10). New York: Academic Press, 243–335.

Ryalls, J. (1996). *A Basic Introduction to Speech Perception.* San Diego: Singular Publishing Group.

Scherer, R.C. (1995). Laryngeal function during phonation. In Rubin, J.S., Korovin, G., Sataloff, R.T., & Gold, W.J. (Eds.). *Diagnosis and Treatment of Voice Disorders.* NY: Igakushoin Medical Publishers, 86–104.

Schouten, M.E.H. (1992). *The Auditory Processing of Speech.* Berlin: Mouton de Gruyter.

Seikel, J.A., King, D.W., & Drumright, D.G. (1997). *Anatomy and Physiology for Speech, Language, and Hearing.* San Diego: Singular Publishing Group.

Sendlemeier, W.G. (1995). Feature, phoneme, syllable, or word: How is speech mentally represented? *Phonetica* 52, 131–143.

Sinott, J., Beecher, M., Moody, D., & Stebbins, W. (1976). Speech sound discrimination by monkeys and humans. *Journal of the Acoustical Society of America* 60, 687–695.

Speaks, C.E. (1996). *Introduction to Sound: Acoustics for the Hearing and Speech Sciences.* (Second Edition). San Diego: Singular Publishing Group.

Stevens, K.N. (1989). On the quantal nature of speech. *Journal of Phonetics* 17, 3–46.

Stevens, K.N. (1998). *Acoustic Phonetics.* Cambridge, MA: MIT Press.

Stevens, K.N., & House, A.S. (1955). Development of a quantitative description of vowel articulation. *Journal of the Acoustical Society of America* 27, 484–493.

Stevens, K. N., & House, A. S. (1961). An acoustical theory of vowel production and some of its implications. *Journal of Speech & Hearing Research,* 4, 303–320.

Stone, M. (1990). A three-dimensional model of tongue movement based on ultrasound and x-ray beam data. *Journal of the Acoustical Society of America* 85, 2207–2217.

Strange, W. (1989). Dynamic specification of coarticulated vowels spoken in sentence context. *Journal of the Acoustical Society of America* 85, 2135–2153.

Studdert-Kennedy, M. (1976). Speech perception. In Lass, N.J. (Ed.), *Contemporary Issues in Experimental Phonetics.* New York: Academic Press, 243–293.

Titze, I.R. (1994). *Principles of Voice Production.* Boston: Allyn & Bacon.

Tortora, G., & Anagnostakos, N.P. (1990). *Principles of Anatomy and Physiology.* New York: HarperCollins.

Warren, D.W. (1996). Regulation of speech aerodynamics. In Lass, N.J. (Ed.), *Principles of Experimental Phonetics.* St. Louis: Mosby, 46–92.

Warren, R.M. (1996). Auditory illusions and perceptual processing of speech. In N.J. Lass (Ed.), *Principles of Experimental Phonetics.* St. Louis: Mosby, 435–466.

Waters, R., & Wilson, W.A. (1976). Speech perception by rhesus monkeys: The voicing distinction in synthesized labial and velar stop consonants. *Perception and Psychophysics* 19, 285–289.

Watson, C.S., Qiu, E.W., Chamberlain, M.M., & Li, X. (1996). Auditory and visual speech perception: Confirmation of a modality–independent source of individual differences in speech recognition. *Journal of the Acoustical Society of America* 100, 1153–1162.

Werker, J.F., & Polka, L. (1993). Developmental changes in speech perception: New challenges and new directions. *Journal of Phonetics* 21, 83–101.

Werker, J.F., & Tees, R.C. (1984). Cross-language speech perception: Evidence for perceptual reorganization during the first year of life. *Infant Behavior and Development* 7, 49–63.

Yost, W.A. (1994). *Fundamentals of Hearing: An Introduction.* (Third Edition). San Diego: Academic Press.

Zemlin, W.R. (1998). *Speech and Hearing Science: Anatomy and Physiology.* (Fourth Edition). Boston: Allyn & Bacon.

Index

Numbers followed by the letter *f* indicate figures; numbers followed by the letter *t* indicate tables.

Abducens nerve, 63*f*
Abductor muscles, 30
Absorption, defined, 89, 185
Accessory nerve, 63*f*
Acoustic filtering, 123, 156*f*
Acoustic nerve, 52, 62, 185, 186
Acoustic-phonetic invariance, 168
Acoustical model of speech production, 6–8, 7*f*, 120–130, 130*f*
  compared to anatomical model, 7–8, 8*f*
Acoustical resonance, 111–121, 185
Acoustics
  fundamentals of, 71–107
  resonance and, 109–126
  of speech production, 129–159
Active theories of speech perception, 174
Adam's apple, 27
Adductor muscles, 30
Aditus laryngis, 35, 36*f*
Air
  compression of, 73*f*
  rarefaction of, 73*f*
Alveolar ridge, 37*f*, 40, 40*f*, 185
Amplitude, 78–80, 79*f*
  defined, 78, 185
Analysis, brain involvement in, 56, 57
Analysis-by-synthesis theory of speech perception, 177–178
Anatomical model of speech production, 3*f*, 4
  compared to acoustical model, 7–8, 8*f*
Anterior faucial pillars, 40*f*
Antihelix, 45*f*
Antitragus, 45*f*
Aperiodicity, 95, 96*f*, 97, 97*f*, 185
Articulation, 5–7, 6*f*, 15
  anatomy of, 37–43, 37*f*
  defined, 185
Aryepiglottic folds, 35, 36*f*
Arytenoid cartilages, 27, 27*f*, 28, 28*f*, 185
Arytenoid muscles, 31*f*, 33, 33*f*
Attic, 47, 47*f*
Audition. *See* Hearing
Auditory cortex, 52, 185
Auditory illusions, 173
Auditory induction, 173
Auditory radiations, 52
Auditory tube, 43*f*, 47, 47*f*
Auricle, 44, 44*f*, 45*f*, 46, 185
Autonomic nervous system, 65–67, 66*f*, 185
Autonomous theories of speech perception, 174

Back vowels, spectrograms of, 164*f*
Bandwidth, defined, 121, 185
Basilar membrane, 50, 50*f*
Bel, defined, 104
Bell, Alexander Graham, 104
Bernoulli's principle, 139–140, 141*f*, 185
Body language, 12
Bottom-up theories of speech perception, 174
Brain, 54, 185
  anatomy of, 54*f*, 61*f*
  auditory cortex of, 52

Brain *(continued)*
functions of, 55*f*, 57*f*
hemispheres of, 55–59, 55*f*
and speech, 10
subhemispheric structures of, 60–61
Brainstem, 61
Breastbone, 17
Breathing, medullar involvement in, 61*f*
Broca's area, 56*f*, 58, 59, 185
Bronchi, 24, 24*f*
Bronchioles, 24, 24*f*

Categorical perception, 171, 176, 185
Cavity resonance, 111, 185
demonstration of, 112, 113*f*, 114–115
mathematics of, 116–120
pressure curve for, 115, 115*f*
velocity curve for, 114, 114*f*
Central auditory pathway, 51*f*
Central fissure, 56, 56*f*, 186
Central nervous system, 54–61, 54*f*, 186
and speech production, 10, 130
Cerebellum, 54*f*, 56*f*, 60, 61*f*, 186
Cerebral hemispheres, 55–59, 55*f*, 186
Cerumen, 46, 186
Cervical vertebrae, 16, 18*f*
Coarticulation, 38, 169, 186
Coccygeal vertebrae, 16, 18*f*
Coccyx, 16
Cochlea, 44*f*, 48*f*, 49, 49*f*, 50*f*, 186
connection to temporal lobe, 52*f*
function of, 53–54, 62
Cochlear nerve, 50*f*, 51, 186
Cochlear nucleus, 51
Cohort theory of speech perception, 180–182, 181*f*
Complex sounds, 92–93, 92*f*, 101*f*, 186
Compression
of air, 73*f*
defined, 77
Concha, 45*f*
Condensation, defined, 77, 186
Cone of light, 46*f*, 47
Consonants
formant transitions in, 165*f*
perception of, 164–165
Continuants, 140
Contralateral connections, defined, 173
Conus elasticus, 35, 36*f*
Corniculate cartilages, 27*f*, 28, 28*f*, 186
Corpus callosum, 60, 61*f*, 186
Cortex, cerebral, 55, 56*f*, 186
Costal cartilage, 17
Cranial nerves, 63*f*, 186
Cricoarytenoid muscles, 30–31, 31*f*, 32*f*
Cricoid cartilage, 25, 27*f*, 28*f*, 186
Cricothyroid muscles, 31*f*, 33, 34*f*
Cricotracheal membrane, 35, 35*f*
Cycles, 81, 81*f*, 186

Damped resonator
defined, 121
frequency response of, 121*f*
Decibel, defined, 80, 187
described, 102–106
mathematics of, 103–106
Declarative statement, intonation pattern of, 166*f*, 167
Deflection, 89, 190
Density, 85*f*
effect on velocity, 85, 86*f*
Diaphragm, 18, 20*f*, 186
Dichotic listening, 172–173, 186
Digastricus muscles, 29, 29*f*
Diphthongs, defined, 162
perception of, 163
spectrograms of, 164*f*
Direct current (dc) air flow, 130, 131*f*
Duplex perception, 177
Duration, 167

Ear
anatomy of, 53*f*
inner, 48–54
middle, 47–48
outer, 43–47
Eardrum, 44, 44*f*, 45*f*, 186
Earwax, 46
Effective length, defined, 112
Eighth cranial nerve, 50*f*, 51, 52, 186

Elasticity, defined, 72, 186
Emotions
  brain involvement in, 59
  nervous system involvement in, 66–67
Epiglottis, 28, 28*f*, 35*f*, 36, 186
Epitympanic recess, 47, 47*f*
Esophagus, 43*f*
Eustachian tube, 43*f*, 47, 47*f*
Exhalation, muscles of, 4, 22–23
Expansion, defined, 77
Exponent, defined, 103, 186
External auditory meatus, 44, 44*f*, 45*f*, 46
External oblique muscle, 22, 23*f*
Extrinsic laryngeal membranes, 35
Extrinsic laryngeal muscles, 29–30, 30*f*

Facial nerve, 62, 63*f*
False ribs, 17, 19*f*
False vocal folds, 35, 36*f*, 186
Faucial pillars, 40*f*
Fenestra rotunda, 44*f*, 47*f*, 48*f*, 49
Fenestra vestibuli, 44*f*, 47*f*, 48, 48*f*, 49
Fissure of Rolando, 56, 56*f*, 186
Fissure of Sylvius, 56, 56*f*, 188
Floating ribs, 17, 19*f*
Formant transitions, 162, 165*f*, 187
Formants, defined, 162, 187
Fossa triangularis, 45*f*
Fourier, J.B., 93
Fourier analysis, 94*f*, 95, 187
Frequency, 82, 82*f*
  relation to period, 82, 83*f*
  relation to wavelength, 83–84, 84*f*
  vocal, 135–137, 137*f*, 167*f*
Frequency response curve, 120*f*, 121, 187
Frequency theories of hearing, 53–54, 187
Fricatives
  defined, 187
  production of, 133, 133*f*
  turbulent modulation and, 140
Front vowels, spectrograms of, 164*f*
Frontal lobe, 54*f*, 56*f*, 57, 187
Frontal pole, 57
Fundamental frequency, 187
Fundamental tone, defined, 97

Generative modular system, 130, 187
Geniohyoid muscles, 29, 29*f*
Gestures, as communication, 12
Glossopharyngeal nerve, 62–64, 63*f*
Glottis, 27*f*, 31, 35, 36*f*, 187
  and vocal frequency, 136*f*
  and vocal intensity, 138, 139*f*
Gum ridge, 40, 185

Hard palate, 37*f*, 40, 40*f*, 187
Harmonics, 99, 187
Hearing, 9, 43
  anatomy of, 44–53, 44*f*
  brain involvement in, 59
  and listening, 10
  nervous system involvement in, 66–67
  theories of, 53–54
Hearing level, 106
Hearing threshold level (HTL), 106, 187
Helix, 45*f*
Hemispheres, of brain, 55–59, 55*f*
Hertz, Heinrich, 82
Heschl's gyrus, 52, 56*f*, 59, 62, 187
Homeostasis, defined, 66, 187
Homunculus, 57–58, 57*f*, 187
Hyoepiglottic ligament, 35, 35*f*
Hyoid bone, 25, 26*f*, 27*f*, 29*f*, 35*f*, 188
Hyothyroid ligaments, 35
Hyothyroid membrane, 25, 26*f*, 35
Hypoglossal nerve, 62, 63*f*, 64
Hypothalamus, 60, 61*f*, 188

Illusions, auditory, 173
Incus, 44*f*, 47*f*, 48
Induction, auditory, 173
Inertia, defined, 72, 188
Inferior colliculus, 51
Inferior cornua, 27, 27*f*
Infrahyoid muscles, 29–30, 30*f*, 188
Inhalation, muscles of, 4, 18, 21
Inner ear, 48–54, 48*f*, 188
Insula, 59, 188
Intensity, vocal, 137–140, 167
Intensity level, 104–105, 188
Interactive activation, 178

Interactive theories of speech perception, 174
Intercostal muscles, 21, 21*f*, 22*f*
Interference, 88–90
Internal oblique muscle, 22, 23*f*
Interrogative statement, intonation pattern of, 166*f*, 167
Intonation, 167, 188
Intonation patterns, 166*f*
Intrinsic laryngeal membranes, 35
Intrinsic laryngeal muscles, 30–34, 31*f*
Invariance, 168
Inverse square law, 88–89, 88*f*, 188
Ipsilateral connections, defined, 173
Island of Reil, 59, 188

Juncture, 167

Language
  brain involvement in, 57, 59
  components of, 8
  development of, 8–9
  thinking and, 11
Language concepts, 9*f*, 10
  storage of, 10
Laryngeal muscles, 29–34
Laryngopharynx, 42*f*, 43*f*
Larynx, 5, 7, 24, 24*f*, 188
  anatomy of, 25, 25*f*, 26*f*, 27–28, 35, 36*f*
  cartilages of, 27*f*, 28*f*
  membranes of, 26*f*, 35
  role in phonation, 24–25
Lateral cricoarytenoid muscles, 31, 31*f*, 32*f*
Lateral fissure, 56, 56*f*, 188
Lateral hyothyroid ligaments, 35
Lateral ligament of inner ear, 47*f*
Lexicon, 8
Linearity, 168
Lingual dorsum, 40*f*
Lips, 37*f*, 38, 188
  musculature of, 39*f*
Listening
  defined, 10–11
  dichotic, 172–173
Lobe, 45*f*
Lobule, 45*f*
Logarithms
  base–10, 103*t*
  defined, 103, 188
Logogen theory of speech perception, 178–179, 179*f*
Longitudinal waves, 86, 87*f*, 188
Lumbar vertebrae, 16, 18*f*
Lungs, 4, 188
  structure of, 24, 24*f*

Malleus, 44*f*, 46*f*, 47, 47*f*, 48, 188
Mandible, 25, 26*f*
Manubrium, 46*f*, 47
Mass, defined, 72, 188
Mastoid air cells, 47*f*
Medial geniculate body, 52
Medulla, 54*f*, 56*f*, 61, 61*f*, 188
Memory, brain involvement in, 57
Midbrain, 54*f*, 56*f*, 61*f*
Middle ear, 47–48, 47*f*, 188
Modulatory system, 130
Modules, defined, 177
Molecules, 71, 189
Motor activity
  cerebellar involvement in, 60
  cerebral involvement in, 57
Motor cortex, 56*f*, 57, 189
Motor strip, 57
Motor theory of speech perception, 176–177, 175*f*
  revised, 177
Mylohyoid muscles, 29, 29*f*
Myoelastic-aerodynamic theory of phonation, 138, 189

Nasal cavity, 5
Nasal coupling, 145, 146*f*, 147*f*
Nasopharynx, 42*f*, 43*f*
Natural frequency, 109
  defined, 112, 189
Nervous system
  autonomic, 65–67
  central 54–61
  parasympathetic, 66, 189
  peripheral, 62–65

sympathetic, 66, 191
Nodes, defined, 177
Noise, defined, 142, 189
Nucleus of lateral lemniscus, 51

Oblique arytenoid muscles, 31*f*, 33, 33*f*
Occipital lobe, 54*f*, 56*f*, 58, 189
Occipital pole, 56*f*, 58
Oculomotor nerve, 63*f*
Olfactory nerve, 63*f*
Omohyoid muscles, 29, 30*f*
Optic nerve, 63*f*, 64
Oral cavity, 4, 5, 7
Oral-nasal coupling, 145, 146*f*
Orbicularis oris muscle, 38, 39*f*, 62
Organ of Corti, 50, 50*f*, 53, 189
Oropharynx, 42*f*, 43*f*
Ossicles, 48
Ossicular chain, 48
Out of phase, 95, 96*f*
Outer ear, 44, 46–47, 189
Oval window, 44*f*, 47*f*, 48, 48*f*, 49

Parasympathetic nervous system, 66, 189
Parietal lobe, 54*f*, 56*f*, 58, 189
Pars flaccida, 46, 46*f*
Pars tensa, 46, 46*f*
Particle, 71
Particle displacement, 76, 77*f*
Particle pressure, 75–76, 77*f*
Particle velocity, 76, 77*f*
Passive theories of speech perception, 174
Pattern Playback, 161, 163*f*, 189
Peak amplitude, 78, 80*f*, 190
Peak-to-peak amplitude, 79, 80*f*, 189
Pectoral girdle, 17*f*, 20*f*
Pectoralis major, 21, 21*f*
Pectoralis minor, 21, 21*f*
Pelvic region, 17*f*
Perceptual constancy (invariance), 172
Period, 81, 81*f*, 189
relation to frequency, 82, 83*f*
Periodicity, 93*f*
defined, 93, 189
Peripheral nervous system, 62–65, 63*f*
Pharynx, 5, 37*f*, 41–43, 42*f*, 43*f*, 62–63, 189
nerve control over, 64
Phase, 95, 95*f*, 189
Phonation, 5, 5*f*, 15, 189
air flow in, 74*f*
myoelastic-aerodynamic theory of, 138, 189
Phonatory cycle, 94*f*
Phonemes, 5, 8
characteristics of, 169
defined, 168
Phonemic restoration effect, 173
Pinna, 44, 44*f*, 45*f*
Pitch, defined, 82, 189
Pituitary gland, 61
Place theories of hearing, 53, 189
Plane wave, defined, 89, 190
Plane wave excitation, 150
Planum temporale, 56*f*, 59, 190
Plosives, 190
production of, 133, 134*f*
stepwise modulation and, 143–145, 144*f*
Pons, 54*f*, 56*f*, 61*f*
Posterior cricoarytenoid muscles, 30–31, 31*f*, 32*f*
Posterior faucial pillars, 40*f*
Power scale, 104
Prosodic features, 167, 190
Pure tone
defined, 75, 190
generation of, 110*f*, 111
spectrum of, 91*f*

Quadrangular membranes, 35, 36*f*
Quantity timing, 167, 190
Question, intonation pattern of, 166*f*, 167

Rarefaction, of air, 73*f*
defined, 77, 186
Reading
development of, 10
educational role in, 11
Recognition, threshold of, 179
Rectus abdominis, 22, 23*f*
Reflection, defined, 89, 190

Refraction, 90*f*
  defined, 89, 190
Relaxer muscles, 30
Resonance, 109, 190
  acoustical, 111–121
  mathematics of, 116–120, 150–152
  physics of, 112–115
  and speech, 145–157
  vocal tract length and, 155–156
Resonance curve, 121
Resonant frequencies, 109
  defined, 112
  formula for, 150–152
Respiration, 15, 190
  muscles of, 18, 21–24
  skeletal framework for, 16–18, 17*f*
  in speech production, 4*f*
Rest position, defined, 114
Reverberation, 89, 89*f*, 190
Revised motor theory of speech perception, 177
Rib cage, 17–18
Ribs, 17, 17*f*, 19*f*, 20*f*, 190
Right ear advantage, 172
Round window, 44*f*, 47*f*, 48*f*, 49
Rugae, 40*f*

Sacral vertebrae, 16, 18*f*
Scala media, 50, 50*f*
Scala vestibuli, 49
Schwa vowel, 7, 123, 124*f*
  vocal tract shape for, 149*f*
Segmentation, 168
Semicircular canals, 44*f*, 48–49, 48*f*, 190
Sensation, brain involvement in, 58
Senses
  and language development, 9–10
  overload of, 10
Sensory cortex, 56*f*, 58, 190
Shoulder girdle, 17*f*, 20*f*
Simple harmonic motion, 74–78, 76*f*, 190
Sine wave (curve), 75, 78*f*, 190
Sinusoidal motion, 74–78, 76*f*, 190
Smell, brain involvement in, 59
Soft palate, 40*f*, 41, 41*f*, 42*f*, 190
Sound
  complex, 92–93, 92*f*, 101*f*, 186
  defined, 71, 190
  interference and, 88–90
  propagation of, 88
  spectrum of, 91, 91*f*, 97*f*, 98*f*, 191
  transmission of, 71–72
  turbulent modulation of, 140–142, 143*f*
  velocity of, 85
  vibratory modulation of, 133–140
  visualization of, 100–102
Sound pressure level (SPL), 105–106, 190
Sound ray, defined, 89
Sound spectrograph, 101, 102*f*, 161
Sound waves, longitudinal nature of, 87
Sounding-board effect, 111, 112*f*, 190
Source-filter theory, 129
Spatial relationships, brain involvement in, 56
Speaker-listener model of speech production, 2–4, 2*f*, 3*f*
Spectral envelope, 99, 100*f*, 190
Spectrograms, 102, 103*f*, 162*f*, 163*f*, 191
Spectrograph, 191
Spectrum, of sound, 91, 91*f*, 97*f*, 98*f*, 191
Speech
  cerebellar involvement in, 60
  cerebral involvement in, 10, 52, 56*f*, 59, 62
  energy source for, 130–131
  language concepts and, 11
  minimal unit of perceptual analysis of, 170
  normalization of, 169–170
  perception of, 170–182
  pitch changes in, 147
  production of, 1–8, 16*f*, 168–172
  recognition of, 168
  resonance source for, 145–157
  sound sources for, 131–144
  vocal tract length and, 152–153
Speech and hearing centers, of brain, 56*f*
Speech-language model, 9*f*
Speech perception
  active theories of, 174
  analysis-by-synthesis theory of, 177–178
  autonomous theories of, 174

bottom-up theories of, 174
cohort theory of, 180–182, 181*f*
interactive theories of, 174
logogen theory of, 178–179, 179*f*
motor theory of, 175–176, 175*f*
passive theories of, 174
revised motor theory of, 177
top-down theories of, 174
TRACE model of, 178, 179*f*
Speech production, 16*f*, 168–172
models of, 1–8
Spinal column, 16, 17*f*, 18*f*
Spinal cord, 54, 54*f*, 191
Spinal nerves, 64–65, 65*f*, 191
Spring-mass model, 72*f*
Stapedius muscle, 47*f*
Stapes, 44*f*, 47*f*, 48
Stepwise modulation, 143–145, 144*f*
spectrum and waveform for, 145*f*
Sternohyoid muscles, 29, 30*f*
Sternum, 17, 17*f*, 20*f*, 191
Stops (plosives), 190
perception of, 164
production of, 133, 134*f*
stepwise modulation and, 143–145, 144*f*
Stress, 167, 191
Stylohyoid muscles, 29, 29*f*
Subglottic air pressure, 137, 140*f*, 191
Subglottic portion of larynx, 35, 36*f*, 191
Superior cornua, 27*f*
Superior ligament of inner ear, 47*f*
Superior olivary complex, 51
Suprahyoid muscles, 29, 29*f*
Suprasegmental features, 167
Sympathetic nervous system, 66, 191
Sympathetic vibration, 109, 110*f*, 111, 191
Syntax, 8

Tectorial membrane, 51
Teeth, 38, 39*f*, 40, 191
Tegmen tympani, 47*f*
Temperature, effect on velocity, 85–86
Temporal lobe, 54*f*, 56*f*, 58, 59, 191
Temporal pole, 56*f*, 59
Tensor muscles, 30
Thalamus (thalami), 52, 60, 61*f*, 191
Thinking, 11
cerebral involvement in, 57
Thoracic cavity, 16, 191
Thoracic vertebrae, 16, 18*f*, 19*f*
Thorax, 4
Threshold of recognition, 179
Thyroarytenoid muscles, 33
Thyrohyoid membrane, 35
Thyrohyoid muscles, 30, 30*f*
Thyroid cartilage, 27, 27*f*, 28*f*, 191
Thyroid protuberance, 27*f*
Thyromuscularis muscles, 31*f*, 33, 34*f*
Thyrovocalis muscles, 31*f*, 33, 34*f*
Tongue, 37–38, 38*f*, 191
nerve control of, 64
Top-down theories of speech perception, 174
TRACE model of speech perception, 178, 179*f*
Trachea, 24, 24*f*, 26*f*, 35, 43*f*, 191
Tragus, 45*f*
Transverse arytenoid muscles, 31*f*, 33, 33*f*
Transverse waves, 86–87, 87*f*, 191
Transversus abdominis, 22, 23*f*
Traveling wave theory, 50, 191–192
Trigeminal nerve, 62, 63*f*
Trochlear nerve, 63*f*
True ribs, 17, 19*f*
True vocal folds, 35, 36*f*, 192
Turbulent modulation, 140–142
spectrum and waveform for, 143*f*
Tympanic antrum, 47*f*
Tympanic membrane, 44, 44*f*, 45*f*, 46–47, 47*f*
Tympanum, 47, 47*f*

Umbo, 46*f*, 47
Undamped resonator
defined, 121
frequency response of, 120*f*
Uvula, 40*f*

Vagus nerve, 62, 63*f*, 64
Vegetative breathing, 61

Velocity, 85, 192
  factors affecting, 85–86
Velopharyngeal closure, 41, 192
Velopharyngeal port mechanism, 41, 41*f*, 42*f*, 192
Velum, 40*f*, 41, 41*f*, 42*f*
Ventricle, laryngeal, 35, 36*f*, 192
Ventricle of Morgagni, 35
Ventricular vocal folds, 35
Verbal transformation effect, 173
Vertebrae, 16
Vertebral column, 16, 17*f*, 18*f*, 192
Vertebrochondral ribs, 17, 19*f*
Vertebrosternal ribs, 17, 19*f*
Vestibular apparatus, 48–49, 192
Vestibular nerve, 50*f*, 51, 192
Vestibule, laryngeal, 35, 36*f*, 192
Vibratory modulation, 133–140, 141*f*, 149, 192
  spectrum and waveform for, 135*f*
Vision, brain involvement in, 58
Visual aids, and learning, 11
Vocal folds, 24, 27*f*, 35, 36*f*
  excursions of, 79*f*, 138*f*
  frequency of vibration of, 5
  stiffness of, 136–137
  vibrations of, 154–155
  and vocal intensity, 138
Vocal frequency, 135–137, 167*f*
  gender differences in, 136–137, 137*f*
Vocal intensity, 137–140, 167
  glottis and, 138, 139*f*, 140*f*
Vocal ligaments, 27*f*, 31*f*, 33, 192
Vocal tract, 74, 75*f*, 113*f*, 142*f*, 192
  cavities of, 146*f*
  excitation of, 154–155
  length of, 152–153, 155–156
  plane wave excitation of, 150*f*
  shape of, 153–154
  and speech, 145, 147–148, 148*f*
  tube model of, 122–126, 148–159, 154*f*
Voiced sound source, 131, 132*f*
Voiceless sound source, 131, 132*f*
  aperiodic modulation provided by, 133*f*
Vowels
  duration of, 168
  formant frequencies of, 124*t*, 156*t*
  formation of, 124–125, 152–153, 154–156
  perception of, 162
  resonant frequency relationships of, 152–153
  spectrograms of, 164*f*
  vocal tract shapes for, 125*f*, 126*f*, 148*f*

Waveforms, 90, 192
  characteristics of, 91–92
  complex, 90*f*, 101*f*
  periodicity of, 93–99
Wavelength, 80*f*
  defined, 80–81, 193
  relation to frequency, 83–84, 84*f*
Wernicke's area, 52, 56*f*, 59, 192
White noise, 99, 100*f*, 192
Writing
  development of, 10
  educational role in, 11–12